Lecture Notes in Mathematics

Edited by A. Dold and B. Eckmann

936

S. M. Khaleelulla

Counterexamples in Topological Vector Spaces

Springer-Verlag
Berlin Heidelberg New York 1982

Author
S.M. Khaleelulla
Department of Mathematics
Faculty of Science, King Abdulaziz University
P.O. Box 9028, Jeddah, Saudi Arabia

AMS Subject Classifications (1980): 46 A 05, 46 A 06, 46 A 07, 46 A 09, 46 A 14, 46 A 25, 46 A 35, 46 A 40, 46 B 05, 46 B 15, 46 B 30, 46 C 05, 46 H 05, 46 J 20

ISBN 3-540-11565-X Springer-Verlag Berlin Heidelberg New York
ISBN 0-387-11565-X Springer-Verlag New York Heidelberg Berlin

Printed in Germany

Printing and binding: Beltz Offsetdruck, Hemsbach/Bergstr.
2141/3140-543210

TO

PROFESSOR GALAL M. EL-SAYYAD

PREFACE

During the last three decades much progress has been made in the field of topological vector spaces. Many generalizations have been introduced; this was, to a certain extent, due to the curiosity of studying topological vector spaces for which a known theorem of Functional analysis can be proved. To justify that a class C_1 of topological vector spaces is a proper generalization of another class C_2 of topological vector spaces, it is necessary to construct an example of a topological vector space belonging to C_1 but not to C_2 ; such an example is called a counterexample. In this book the author has attempted to present such counterexamples in topological vector spaces, ordered topological vector spaces, topological bases and topological algebras.

The author makes no claim to completeness, obviously because of the vastness of the subject. He makes no attempt to give due recognition to the authorship of most of the counterexamples presented in this book.

It is assumed that the reader is familiar with general topology. The reader may refer to B[18] for information about general topology.

To facilitate the reading of this book, some fundamental concepts in vector spaces and ordered vector spaces have been collected in the Chapter called 'Prerequisites'. Thereafter each Chapter begins with an introduction which presents the relevent definitions and statements of theorems and propositions with references where their proofs can be found. For some counterexamples which require long and complicated proofs, only reference has been made to the literature where they are available.

The books and papers are listed separately in the bibliography at the end of the book. Any reference to a book is indicated by writing B[ ] and to a paper by P[ ].

The author would like to express his deep gratitude to Professor T. Husain, McMaster University, Hamilton, Canada, and Dr. I. Tweddle, University of Stirling, Stirling, Scotland, who have given him both moral and material support during the preparation of this book. The author wishes to thank Mr. Mohammed Yousufuddin for typing the manuscript.

The author takes great pleasure in thanking the editors and the staff of Springer's "Lecture Notes in Mathematics" series for their keen interest in the publication of this book.

S.M. Khaleelulla

Department of Mathematics
Faculty of Science
King Abdulaziz University
Jeddah, Saudi Arabia

CONTENTS

PREREQUISITES 1

I TOPOLOGICAL VECTOR SPACES 9

Introduction 9

1. A topology on a vector space, which is not compatible with the vector space structure. 12
2. A topological vector space which is not a locally semi-convex space. 12
3. A locally bounded (and hence a locally semi-convex) space which is not a locally convex space. 13
4. A locally convex space which is not a locally bounded space. 14
5. A locally semi-convex space which is neither locally convex nor locally bounded.
   - A metrizable topological vector space which is not locally bounded. 14
6. A topological vector space on which there exist no non-trivial continuous linear functionals. 14
7. A topological vector space such that no finite-dimensional subspace has a topological complement in it. 15
8. Two closed subspaces of a topological vector space, whose sum is not closed. 16
9. A topological vector space in which the convex envelope of a precompact set is not precompact (not even bounded). 16
10. A bounded linear map from a topological vector space to a topological vector space, which is not continuous. 17

II LOCALLY CONVEX SPACES 18

Introduction 18

1. A locally convex space which is not metrizable. 21

2. A metrizable topological vector space which is not locally convex. 21

3. A sequentially complete locally convex space which is not quasi-complete. 21

4. A quasi-complete locally convex space which is not complete. 22

5. A complete locally convex space which is not B-complete. 22

6. A complete locally convex space which is not metrizable. 23

7. A normed space (and hence a metrizable locally convex space) which is not complete. 24

8. A locally convex space which contains a closed, circled and convex set with no extreme points. 24

9. A topological vector space which contains a compact convex set with no extreme points. 25

10. A weakly compact set in a locally convex space, whose weakly closed envelope is not weakly compact. 25

11. A bounded sequence in a topological vector space, which is not convergent. 26

III SPECIAL CLASSES OF LOCALLY CONVEX SPACES 27

Introduction 27

1. An inner product (a pre-Hilbert) space which is not a Hilbert space. 34

2. A generalized inner product space which is not an inner product space. 34

3. A semi-inner product space which is not an inner product space. 34

4. A generalized semi-inner product space which is neither a semi-inner product space nor a generalized inner product space. 35

5. A Banach space which is not a Hilbert space. 36

6. A Banach space which is not separable. 36

7. A Banach space which is not reflexive. 36

8. A Fréchet space which is not a Banach space. 37

9. A t-polar space which is not B-complete.
- A t-polar space which is not barrelled. 37

10. A barrelled space which is not complete.
- A barrelled space which is not a Fréchet space. 37

11. A barrelled space which is not metrizable.
- A barrelled space which is not a Fréchet space. 38

12. A Baire-like space which is not an unordered Baire-like space. 38

13. A Baire-like space which is not a Baire space.
- A barrelled space which is not a Baire space. 38

14. A barrelled bornological space which is not the inductive limit of Banach spaces. 39

15. A bornological space which is not metrizable. 40

16. A bornological space which is not barrelled. 40

17. A barrelled space which is not bornological. 40

18. A quasi-barrelled space which is neither barrelled nor bornological. 41

19. A quasi-M-barrelled space which is not quasi-barrelled. 41

20. A semi-bornological space which is not an S-bornological space (and hence not a bornological space). 41

21. An S-bornological space which is not C-sequential (and hence not a bornological space). 42

22. A C-sequential locally convex space which is not S-bornological (and hence not bornological). 43

23. A Mackey space which is not quasi-barrelled. 44

24. A Mackey space which does not have property (S). 44

25. A Mackey space with property (S) but without property (C). 44

26. A semi-reflexive space which is not reflexive.
– A Mackey and semi-reflexive space which is not reflexive.
– A semi-reflexive space which is not quasi-barrelled.

- A complete locally convex space which is not quasi-barrelled.
- A topological projective limit of barrelled spaces, which is not quasi-barrelled. 45

27. A barrelled space which is not a Montel space. 45

28. A reflexive space which is not a Montel space. 46

29. A Fréchet space which is not a Schwartz space. 46

30. A Schwartz space which is not a Montel space. 46

31. A Montel space which is not separable. 47

32. A Montel space (and hence a reflexive locally convex space) which is not complete.
- A Montel (and hence barrelled) space which is not a Frechet space. 48

33. A distinguished space which is not semi-reflexive. 49

34. A Fréchet space which is not distinguished.
- A barrelled space whose strong dual is not barrelled (not even quasi-barrelled).
- A bornological space whose strong dual is not bornological. 49

35. A distinguished space whose strong dual is not separable. 50

36. A distinguished space whose strong dual is not metrizable. 50

37. A distinguished space which is not quasi-barrelled.
- A semi-reflexive space which is not quasi-barrelled
- A Mackey space which is not quasi-barrelled.
- A semi-reflexive space whose strong dual is not semi-reflexive. 51

38. A bornological space whose strong bidual is not bornological. 51

39. An (LB)-space which is not quasi-complete. 52

40. A locally convex space which is not reflexive (not even semi-reflexive) but its strong dual is reflexive. 53

41. A countably barrelled space which is not barrelled.

– A countably quasi-barrelled space which is not quasi-barrelled.
– A locally convex space C(X) of continuous functions, which is not a Mackey space.
– A complete locally convex space which is not barrelled. 53

42. A locally convex space C(X) of continuous functions which is not countably barrelled. 54

43. A semi-reflexive countably barrelled space which is not barrelled. 54

44. A countably quasi-barrelled (and hence σ-quasi-barrelled) space which is not σ-barrelled.
– A countably quasi-barrelled space which is not countably barrelled. 55

45. A σ-barrelled space which is not a Mackey space. 55

45(a). A σ-barrelled space which is not countably quasi-barrelled (and hence not countably barrelled). 57

46. A Mackey space which is not σ-quasi-barrelled. 57

47. A locally convex space which has property (C), but is not σ-barrelled. 58

48. A sequentially barrelled space which is not σ-quasi-barrelled (and hence not σ-barrelled).
– A Mackey space which is sequentially barrelled but not σ-quasi-barrelled.
– A separable sequentially barrelled space which is not barrelled.
– A sequentially barrelled space which has property (S) but not property (C). 58

49. A sequentially barrelled space which does not have property (S).
– A sequentially barrelled space which is not σ-barrelled. 59

50. A quasi-complete locally convex space which is not sequentially barrelled. 59

51. A (DF)-space which is not countably barrelled. 60

52. A (DF)-space which is not quasi-barrelled. 60

53. A quasibarrelled (DF)-space which is not bornological. 60

54. A locally topological space which is neither a bornological space nor a (DF) space. 61

55. A k-quasi-barrelled space which is not k-barrelled. 62

56. An H-space which is not a distinguished space. 62

57. An H-space which is not metrizable. 63

58. An H-space whose strong dual is not separable. 63

OPEN PROBLEMS 63

IV SPECIAL CLASSES OF TOPOLOGICAL VECTOR SPACES 65

Introduction 65

1. A topological vector space in which the filter condition holds but not the closed neighbourhood condition 68

2. An N-S space which is not an L-W space. 69

3. A locally convex space C(X) of continuous functions, which is barrelled and bornological but not W-barrelled. 69

4. An ultrabarrel which is not convex and which does not have a defining sequence of convex sets. 69

5. An ultrabarrelled space which is not barrelled. 69

6. A barrelled space which is not ultrabarrelled. 70

7. An u^{00}-compact set which is not u-compact. 70

8. An ultrabarrelled space which is not non-meagre. 71

9. An ultrabornological space which is not bornological. 71

10. A bornological space which is not ultrabornological. 72

11. An ultrabornological space which is not ultrabarrelled. 72

12. An ultrabarrelled space which is not ultrabornological. 72

13. A quasi-ultrabarrelled space which is neither ultrabarrelled nor ultrabornological. 73

14. A countably quasi-ultrabarrelled space which is not countably ultrabarrelled. 73

15. A countably ultrabarrelled space which is not ultrabarrelled.
– A countably quasi-ultrabarrelled space which is not quasi-ultrabarrelled. 73

16. A countably barrelled space which is not countably ultrabarrelled.
– A countably quasi-barrelled space which is not countably quasi-ultrabarrelled. 73

17. A k-quasi-ultrabarrelled space which is not k-ultrabarrelled. 74

18. A hyperbarrelled space which is not hyperbornological. 75

19. A hyperbornological space which is not hyperbarrelled. 75

20. A quasi-hyperbarrelled space which is neither hyperbarrelled nor hyperbornological. 75

21. An $\mathscr{Y}$-quasi-hyperbarrelled space which is not $\mathscr{Y}$-hyperbarrelled. 75

22. A barrelled space which is not $\mathscr{Y}_0$-hyperbarrelled. 75

V ORDERED TOPOLOGICAL VECTOR SPACES 77

Introduction 77

1. An ordered topological vector space with generating cone which does not give open decomposition. 85

2. An ordered topological vector space with normal cone but with a (topologically) bounded set which is not order-bounded. 85

3. A cone in a topological vector space, which is not normal. 86

4. An ordered topological vector space in which order bounded sets are bounded but the cone is not normal. 86

5. A cone in a topological vector space, which has no interior points. 87

6. An element of a cone in a vector space, which is an interior point for one topology but not for another topology. 87

7. A cone in a locally convex space, which is not a b-cone. 88

8. A base of a cone in a topological vector space, which is not closed. 89

9. An ordered normed space which is not an order-unit normed space though its dual is a base normed space. 89

10. An ordered topological vector space which is complete but not order-complete. 90

11. An ordered topological vector space which is order-complete but not complete. 90

12. An ordered topological vector space which is complete and order-complete but not boundedly order-complete. 90

13. An order-continuous linear functional on an ordered topological vector space, which is not continuous. 91

14. A continuous linear operator on an ordered topological vector space, which is not sequentially order-continuous. 91

15. A positive linear functional on an ordered topological vector space, which is not continuous. 93

16. An ordered topological vector space on which there exist no non-zero positive linear functionals. 93

17. A topological vector lattice which has no non-zero real lattice homomorphisms. 94

18. A topological vector space with lattice ordering in which the map $x \to x^+$ is continuous for all x but not uniformly continuous. 94

19. An ordered locally convex space with a positive weakly convergent sequence which is not convergent. 95

20. An M-space which is not normable. 96

21. A pseudo-M-space which is not an M-space. 96

22. A topological vector lattice which is not a pseudo-M-space. 97

23. The topology of a bornological locally convex lattice which is not an order bound topology.
– A quasi-barrelled locally convex lattice which is not order-quasi-barrelled. 97

24. An order -quasi-barrelled vector lattice which is not barrelled. 98

25. A C.O.Q. vector lattice which is not order-quasi-barrelled.
– An order-(DF)-vector lattice which is not order-quasi-barrelled. 98

26. A C.O.Q. vector lattice which is not countably barrelled.
– An order-(DF)-vector lattice which is not countably barrelled. 99

27. A countably quasi-barrelled locally convex lattice which is not a C.O.Q. vector lattice. 99

28. An order-quasi-barrelled (and hence a C.O.Q.) vector lattice which is not an order-(DF)-vector lattice. 100

29. An O.Q.U. vector lattice which is not ultra-barrelled. 100

30. A quasi-ultrabarrelled topological vector lattice which is not an O.Q.U. vector lattice. 100

31. An order-quasi-barrelled vector lattice which is not an O.Q.U. vector lattice. 100

32. A countably O.Q.U. vector lattice which is not countably ultrabarrelled. 101

33. A countably quasi-ultrabarrelled topological vector lattice which is not a countably O.Q.U. vector lattice. 101

34. A C.O.Q. vector lattice which is not a countably O.Q.U. vector lattice. 101

VI HEREDITARY PROPERTIES 103

Introduction 103

1. A closed subspace of a reflexive space, which is not reflexive.

– A closed subspace of a Montel space, which is not Montel. 104

2. A closed subspace of a bornological space, which is not bornological. 104

3. An infinite countable codimensional subspace of a bornological space, which is not quasibarrelled.
– An infinite countable codimensional subspace of a bornological space, which is not bornological. 104

4. A closed subspace of a barrelled space, which is not countably quasi-barrelled.
– A closed subspace of a barrelled (quasi-barrelled, countably barrelled or countably quasi-barrelled) space, which is not a barrelled (quasi-barrelled, countably barrelled or countably quasi-barrelled) space. 105

5. A dense uncountable dimensional subspace of a barrelled space, which is not barrelled. 105

6. A closed subspace of a (DF)-space which is not a (DF)-space.
– A closed subspace of a barrelled (quasi-barrelled, bornological) space, which is not barrelled (quasi-barrelled, bornological).
– A closed subspace of a Montel space, which is not Montel.
– A closed subspace of a countably quasi-barrelled (countably barrelled) space which is not countably quasi-barrelled (countably barrelled). 106

7. An infinite countable codimensional subspace of a quasi-barrelled (DF) space, which is not a (DF) space. 107

8. A closed subspace of a hyperbarrelled space, which is not hyperbarrelled.
– A closed subspace of a quasi-hyperbarrelled ($\mathcal{N}$-hyperbarrelled, $\mathcal{N}$-quasi-hyperbarrelled) space which is not quasi-hyperbarrelled ($\mathcal{N}$-hyperbarrelled, $\mathcal{N}$-quasi-hyperbarrelled). 107

9. A closed subspace of an ultrabarrelled space, which is not countably quasi-ultrabarrelled. 107

– A closed subspace of an ultrabarrelled (quasi-ultrabarrelled, countably ultrabarrelled, countably quasi-ultrabarrelled) space which is not ultrabarrelled (quasi-ultrabarrelled, countably ultrabarrelled, countably quasi-ultrabarrelled). 107

10. A lattice ideal in an order-quasi-barrelled vector lattice, which is not order-quasi-barrelled.
– A lattice ideal in a C.O.Q. vector lattice, which is not a C.O.Q. vector lattice.
– A lattice ideal in an O.Q.U. vector lattice which is not an O.Q.U. vector lattice.
– A lattice ideal in a countably O.Q.U. vector lattice, which is not a countably O.Q.U. vector lattice. 108

11. A complete locally convex space whose quotient is not sequentially complete.
– A complete (quasi-complete, sequentially complete) space whose quotient is not complete (quasi-complete, sequentially complete). 108

12. A quotient of a Montel space, which is not semi-reflexive.
– A Montel (reflexive, semi-reflexive) space whose quotient is not a Montel (reflexive, semi-reflexive) space. 108

13. A quotient of a Fréchet Montel space, which is not reflexive.
– A Fréchet Montel space whose quotient is not a Montel space.
– A reflexive Fréchet space whose quotient is not reflexive. 108

14. A product of B-complete spaces which is not B-complete. 109

15. An arbitrary direct sum of B-complete spaces, which is not B-complete. 109

VII TOPOLOGICAL BASES 110

Introduction 110

1. A separable Banach space which has no basis. 115

A Banach space with a basis, whose dual space does not have a basis. 115

3. A Banach space which has no unconditional basis. 115

4. A Banach space with a basis which is not unconditional. 116

5. A Banach space with an unconditional basis which is not boundedly complete. 117

6. A Banach space with a basis which is not absolutely convergent. 118

7. A Banach space with a basis which is not a normal basis. 119

8. A Banach space whose dual space has a normal basis which is not a retro-basis. 120

9. A Banach space with a Besselian basis which is not a Hilbertian basis. 120

10. A Banach space with a Hilbertian basis which is not a Besselian basis. 120

11. A Banach space with a basis which is not a monotonic basis. 121

12. A Banach space with a sub-symmetric basis which is not a symmetric basis. 121

13. A Banach space without a sub-symmetric basis. 122

14. An E-complete biorthogonal system in a Banach space, which is not a basis. 122

15. A normed space with a basis which is not a Schauder basis. 122

16. A normed space with a Schauder basis which is neither an (e)-Schauder basis nor a (b)-Schauder basis. 123

17. A Banach space whose dual has a weak* basis but no basis.
— A Banach space whose dual has a weak*-Schauder basis which is not a (Schauder) basis. 124

18. A Banach space whose dual space has a basis which is not a weak* basis. 125

19. A Banach space whose dual space has a weak* basis which is not a weak* Schauder basis. 126

20. A separable locally convex space which has no basis. 128

21. A basis in a locally convex space, which is not a Schauder basis. 128

22. A complete, metrizable and separable (non-locally convex) topological vector space which has no basis. 130

23. A generalized basis in a non-separable Banach space, which is not a Markushevich basis. 130

24. A Markushevich basis in a Fréchet space, which is not a Schauder basis. 130

25. A maximal biorthogonal system in a Fréchet space which is not a generalized basis. 131

26. An extended unconditional basis in a countably barrelled space, which is not an extended unconditional Schauder basis. 132

27. The isomorphism theorem fails if the domain or the range space is not barrelled. 132

28. The isomorphism theorem does not hold for generalized basis even if the domain and the range spaces are complete and barrelled. 133

29. A vector space with two compatible locally convex topologies such that there is a Schauder basis for one topology, which is not a Schauder basis for the other topology. 133

VIII TOPOLOGICAL ALGEBRAS 137

Introduction 137

1. An algebra which cannot be made into a Banach algebra. 142

2. A Banach algebra which has no radical. 142

3. A Banach algebra with a closed ideal which is not an intersection of maximal regular ideals. 142

4. A Banach algebra with an approximate identity

which is not an identity. 143

5. An A^*-algebra which is not a B^*-algebra. 143

6. An A^*-algebra which is not symmetric. 144

7. A Fréchet algebra which is not a Banach algebra.
– A Q-algebra which is not a Banach algebra. 144

8. A Fréchet algebra which is not a locally m-convex algebra.
– A Fréchet algebra which is not a Banach algebra.
– A locally convex algebra which is not a locally m-convex algebra. 145

9. A locally m-convex algebra which is not metrizable.
– A locally m-convex algebra which is a Q-algebra but not a normed algebra. 145

10. A Fréchet algebra which has closed ideals but not closed maximal ideals. 146

11. A Fréchet algebra which does not have the Wiener property.

– A Fréchet algebra which is not a locally m-convex algebra. 147

12. A semi-simple locally m-convex Fréchet algebra which is a projective limit of Banach algebras which are not semi-simple. 147

13. $\mathcal{M}$-singular elements of a locally m-convex Fréchet algebra, which are not topological divisors of zero. 149

14. A locally m-convex Fréchet algebra which has neither topological divisors of zero nor $\mathcal{M}$-singular elements. 150

15. An m-barrelled algebra which is not barrelled. 150

16. A countably m-barrelled algebra which is not m-barrelled. 151

17. A complete p.i.b. algebra which is not a P-algebra. 151

18. A metrizable p.i.b. algebra which is neither a P-algebra nor an m-bornological algebra. 152

19. The Gelfand map which is continuous for a locally convex algebra which is not m-barrelled. 152

20. A GB^*-algebra which is not a locally m-convex algebra. 153

21. A GB^*-algebra on which there are no non-trivial multiplicative linear functionals. 153

22. A Pseudo-complete locally convex algebra which is not sequentially complete. 154

23. An A-convex algebra which is not a locally m-convex algebra. 155

24. A p-normed (locally bounded) algebra which is not a normed algebra. 156

25. A locally m-semi-convex algebra which is not a locally m-convex algebra. 157

OPEN PROBLEMS 157

BIBLIOGRAPHY 158

INDEX 170

PREREQUISITES

VECTOR SPACES AND ORDERED VECTOR SPACES

A nonempty set E is called a vector space over a field $\mathbb{K}$ if

(a) E is an additive abelian group, and

(b) for every $\alpha \varepsilon \mathbb{K}$ and $x \varepsilon E$, there is defined an element αx in E subject to the following conditions:

(b_1) $\alpha(x+y) = \alpha x + \alpha y$

(b_2) $(\alpha+\beta)x = \alpha x + \beta x$

(b_3) $\alpha(\beta x) = (\alpha\beta)\, x$

and (b_4) $1x = x$

for all $\alpha, \beta \,\varepsilon\, \mathbb{K}$, $x,y \,\varepsilon\, E$ and 1 the unit element of $\mathbb{K}$ under multiplication.

If $\mathbb{K}$ is the field $\mathbb{R}(\mathbb{C})$ of real (complex) numbers, the vector space E is called a real (complex) vector space.

Throughout this book, we deal with only real or complex vector spaces, and we use 0 to denote the zero element of $\mathbb{K}$ as well as that of a vector space.

PROPOSITION 1. If E is a vector space over $\mathbb{K}$,

(a) $\alpha 0 = 0$ for all $\alpha \,\varepsilon\, \mathbb{K}$;

(b) $0x = 0$ for $x \varepsilon E$;

(c) $(-\alpha)x = -(\alpha x)$ for $\alpha \,\varepsilon\, \mathbb{K}$, $x \,\varepsilon\, E$;

and (d) $\alpha x = 0$, $x \neq 0$, implies that $\alpha = 0$.

A vector space E with a multiplication (that is, if $x,y \,\varepsilon E$, then $xy, yx \,\varepsilon E$) is called an algebra.

If E is a vector space and F a nonempty subset of E, then F is called a vector subspace (or simply, subspace) of E if, under the operations of E, F itself forms a vector space over the field $\mathbb{K}$. If $x_1, \ldots, x_n \in E$, then $\sum_{i=1}^{n} \alpha_i x_i$, $\alpha_i \in \mathbb{K}$, is called a linear combination of $x_1, \ldots, x_n$. A subset B of a vector space E is called linearly independent if $B \neq \emptyset$ or $\{0\}$ and no element of B is a linear combination of any finite subset of other elements of B. A maximal linearly independent subset of a vector space is called a Hamel basis (or vector basis). Every vector space has a Hamel basis and any two Hamel bases of a vector space have the same cardinal number. The cardinal number of a Hamel basis of a vector space is called its dimension.

If F is a subspace of a vector space E over the field $\mathbb{K}$, the quotient space of E by F is a vector space E/F over $\mathbb{K}$ where, for $x_1 + F, x_2 + F \in E/F$ and $\alpha \in \mathbb{K}$,

(i) $(x_1 + F) + (x_2 + F) = (x_1 + x_2) + F$

and (ii) $\alpha(x_1 + F) = \alpha x_1 + F$.

An arbitrary product $E = \Pi_\alpha E_\alpha$ of vector spaces E_α is a vector space where addition and scalar multiplication are defined as coordinatewise addition and scalar multiplication.

If $\{E_\alpha\}$ is a family of vector spaces and

$F = \sum_{\alpha \in I} E_\alpha = \{x = \{x_\alpha\} ;\ x_\alpha = 0$ for all α except for a finite subset of $I\}$, then F is a vector space, called direct sum of $\{E_\alpha\}$, where addition and multiplication are defined as above.

A map f of a vector space E into another vector space F is said to be linear if

$$f(\alpha x + \beta y) = \alpha f(x) + \beta f(y)$$

for all $x,y \in E$ and $\alpha, \beta \in \mathbb{K}$. If $F = \mathbb{K}$, then f is called a linear functional.

A subset A of a vector space E is said to be (i) circled (or balanced) if $\alpha A \subset A$ for every $\alpha \in \mathbb{K}$ such that $|\alpha| \leq 1$,

(ii) absorbing if for every $x \in E$ there is an $\alpha > 0$ such that $x \in \lambda A$ for all $\lambda \in \mathbb{K}$ with $|\lambda| \geq \alpha$, (iii) convex if $x,y \in A$ and $0 \leq \lambda \leq 1$ imply that $\lambda x + (1 - \lambda)y \in A$, and (iv) semiconvex if $A+A \subset \lambda A$ for some $\lambda > 0$.

Let E be a vector space. A map $p: E \to \mathbb{R}^+$ is called a semi-norm if

(a) $p(x+y) \leq p(x) + p(y)$ for all $x,y \in E$,

and (b) $p(\lambda x) = |\lambda| \; p(x)$ for all $x \in E$ and $\lambda \in \mathbb{K}$.

Clearly $p(0) = 0$. If $p(x) = 0$ implies $x = 0$, then p is called a norm on E and is denoted by $|| \cdot ||$.

p is called a k-semi-norm if (b) is replaced by the following:

(b') $p(\lambda x) = |\lambda|^k \, p(x)$, $0 < k \leq 1$, $x \in E$ and $\lambda \in \mathbb{K}$.

Clearly $p(0) = 0$. If $p(x) = 0$ implies $x = 0$, then p is called a k-norm.

p is called a quasi-semi-norm if (a) is replaced by the following:

(a') There is a number $b \geq 1$ for which $p(x+y) \leq b(p(x) + p(y))$ for all $x,y \in E$.

The smallest value of b for which (a') is satisfied is referred to as the multiplier of p.

Clearly $p(0) = 0$. If $p(x) = 0$ implies $x = 0$, then p is called a quasi-norm.

THEOREM 1. If q is a quasi-semi-norm on a vector space E with multiplier b and $k < \log_{2b}^{2}$, then there is a k-semi-norm p on E equivalent to q.

The inner product $(.\,,.)$ in a vector space E is a map $E \times E \to \mathbb{K}$ satisfying the following conditions:

(a) $(x,x) \geq 0$ for all $x \in E$;

(b) $(x,x) = 0$ iff $x = 0$;

(c) $(x,y) = (y,x)$ for all $x, y \in E$;

(d) $(\lambda x + \mu y, z) = \lambda(x,z) + \mu(y,z)$ for all $x,y \in E$ and $\lambda, \mu \in \mathbb{K}$.

The semi-inner product $[.\,,.]$ in a vector space E is a map $E \times E \to \mathbb{K}$ satisfying the following conditions:

(a_1) $[x+y,z] = [x,z] + [y,z]$, $x,y,z \in E$

(b_2) $[\lambda x,y] = \lambda[x,y]$, $x,y \in E$, $\lambda \in \mathbb{K}$;

(c_3) $[x,x] > 0$ for $x \neq 0$, $x \in E$;

(d_4) $|[x,y]| \leq [x,x]^{\frac{1}{2}} [y,y]^{\frac{1}{2}}$, $x,y \in E$.

All vector spaces in what follows are over the field $\mathbb{R}$ of real numbers.

A vector space E is called an ordered vector space if it is equipped with a reflexive, transitive and antisymmetric relation $\leq$ satisfying the following conditions:

(a) $x \leq y$ implies $x + z \leq y + z$ for all $x,y,z \in E$; and (b) $x \leq y$ implies $\lambda x \leq \lambda y$ for all $x,y \in E$ and $\lambda \in \mathbb{R}, \lambda > 0$.

The set $C = \{x \in E ; x \geq 0\}$ is called the positive cone (or simply, cone) in an ordered vector space E. It satisfies the following conditions:

(a_1) $C + C \subset C$;

(b_2) $\lambda C \subset C$ for $\lambda > 0$;

(c_3) $C \cap (-C) = \{0\}$.

On the other hand, if C is a subset of a vector space E satisfying (a_1), (b_2) and (c_3) then $x \leq y$ if and only if (henceforth abbreviated to iff) $y - x \in C$ defines an order relation $\leq$ on E for which E becomes an ordered vector space with C as positive cone.

We write (E,C) to denote an ordered vector space E with the positive cone C.

A subset C of E satisying (a_1) and (b_1) is called a wedge.

The order-interval between two elements x and y of an ordered vector space is the set $\{t \in E; x \leq t \leq y\}$ which is denoted by $[x,y]$. A subset B of E is said to be order-bounded if there exist x,y in E such that $B \subseteq [x,y]$. A subset A of E is said to be majorized (minorized) if there is an element t in E such that $t \geq a$ (respectively, $t \leq a$) for all $a \in A$. If every pair x,y in A is majorized (minorized), then A is said to be directed $\leq$ (respectively, directed $\geq$).

The positive cone C in an ordered vector space E is generating if $E = C - C$. The positive cone C is generating iff E is directed $\leq$. An element e in (E,C) is called an order-unit if the order interval $[-e,e]$ is absorbing, that is, if $e \varepsilon C$ and for each $x \varepsilon E$ there exists $\alpha > 0$ such that $-\alpha e \leq x \leq \alpha e$. If (E,C) contains an order-unit, then C is generating. A net $\{e_\lambda ; \lambda \varepsilon \Lambda, \leq\}$ in (E,C) is called an approximate order-unit if the following conditions are satisfied:

(i) $e_\lambda \varepsilon C$ for each λ;

(ii) for any pair of elements λ_1, λ_2 in the directed set Λ with $\lambda_1 \leq \lambda_2$, we have $e_{\lambda_1} \leq e_{\lambda_2}$;

(iii) for each x in E there exist $\lambda \varepsilon \Lambda$ and a real number $\alpha > 0$ such that $-\lambda e_\lambda \leq x \leq \alpha e_\lambda$.

Clearly each order-unit is approximate order-unit.

An ordered vector space is called Archimedean (almost Archimedean) if $x \leq 0$ whenever $\alpha x \leq y$ for some $y \varepsilon C$ and all $\alpha > 0$ (respectively, if $x = 0$ whenever $-\lambda y \leq x \leq \alpha y$ for some $y \varepsilon C$ and all $\alpha > 0$). Every Archimedean ordered vector space is almost Archimedean. A set A in an ordered vector space (E,C) is said to be decomposable if for each a in A there exist a_1, a_2 in $A \cap C$ such that $a = \alpha_1 a_1 - \alpha_2 a_2$ for some $\alpha_1, \alpha_2 \geq 0$ with $\alpha_1 + \alpha_2 = 1$. A subset A of an ordered vector space (E,C) is said to be full if

$$A = \{z \varepsilon E ; \ x \leq z \leq y \text{ for } x,y \varepsilon A\} .$$

A subset S of the positive cone C in an ordered vector space is said to be exhausting if for each $x \varepsilon C$, there are $s \varepsilon S$ and $\lambda > 0$ such that $x \leq \lambda s$. A nonempty convex subset B of the positive cone C in an ordered vector space (E,C) is a base for C if each $x \in C$, $x \neq 0$, has a unique representation of the form $x = \lambda b$, $b \in B$, $\lambda > 0$. x is an extremal point of the cone C if each point of the order interval $[0,x]$ is a positive scalar multiple of x.

A linear map T from an ordered vector space (E,C) into another ordered vector space (F,K) is positive if $Tx \in K$ whenever $x \in C$. A linear functional on an ordered vector space (E,C) is positive if $Tx \geq 0$ whenever $x \in C$.

Let (E,C) be an ordered vector space. Let A be a subset of E satisfying the following properties:

(S_1) $x \geq a$ for all $a \in A$;

(S_2) $y \geq x$ whenever $y \geq a$ for all $a \in A$.

Then x is called the supremum of A and is written as $x = \sup(A)$. Dually we can define the infimum of A.

If $\sup\{x,y\}$, written as $x \vee y$, or $\inf\{x,y\}$, written as $x \wedge y$, of every pair x,y of elements of E exists in E, then E is called a vector lattice.

$x^+ = \sup\{x,0\}$, $x^- = (-x)^+$, $|x| = \sup\{x, -x\}$ are respectively called the positive part, negative part and absolute value of x in a vector lattice E. It follows that $x = x^+ - x^-$ so that the positive cone in a vector lattice is always generating.

In a vector lattice (E,C) the following property, called the decomposition property, is always satisfied:

$$[0,x] + [0,y] = [0, x + y], \quad x,y \in C.$$

An ordered vector space (a vector lattice) is order-complete (σ-order complete) if every directed subset D, $\leq$ of

E that is majorized in E has supremum that belongs to E (if the supremum of every countable majorized subset of E exists in E). An order-complete vector space (E,C) is a vector lattice iff C is generating. An ordered vector space (E,C) is order-separable if every subset B of E that has a supremum in E contains a countable subset B_1 such that $\sup(B) = \sup(B_1)$.

A net $\{x_\alpha\}$ in a vector lattice E decreases to $x_0 \in E$ if $x_0 = \inf\{x_\alpha\}$ and $x_\alpha \geq x_\beta$ whenever $\beta \geq \alpha$. A net $\{x_\alpha\}$ in E order-converges to $x_0 \in E$ if $\{x_\alpha\}$ is an order-bounded subset of E and there is a net $\{y_\alpha\}$ that decreases to 0 such that $|x_\alpha - x_0| \leq y_\alpha$ for all α. A linear map T from a vector lattice E into a vector lattice F is order-continuous if the net $\{Tx_\alpha\}$ order-converges to 0 in F whenever $\{x_\alpha\}$ is a net that order-converges to 0 in E. Instead of nets, if sequences are considered, then we call it as sequentially order-continuous.

A linear map from a vector lattice (E,C) into another vector lattice (F,K) is called a lattice homomorphism if it preserves lattice operations $\vee$ and $\wedge$.

A subset A of a vector lattice (E,C) is called solid if $|x| \leq |y|$, $y \in A$ implies $x \in A$. A solid subspace of E is called a lattice ideal. A subspace F of a vector lattice (E,C) is called a sublattice if for every pair $x,y \in E$ the supremum and infimum of x and y in E lies in F. Every lattice ideal is a sublattice. A lattice ideal I in an order-complete vector lattice is called a band in E if I contains the supremum of every subset of I that is majorized in E.

CHAPTER I

TOPOLOGICAL VECTOR SPACES

Introduction

A topological space E which is also a vector space over the field $\mathbb{K}$ of reals or complexes is called a topological vector space if

(TVS1) the map $(x,y) \to x+y$ from $E \times E$ into E is continuous, and

(TVS2) the map $(\lambda, x) \to \lambda x$ from $\mathbb{K} \times E$ into E is continuous.

A topology on a vector space is said to be compatible if the axioms (TVS1) and (TVS2) are satisfied.

REMARK 1. We shall usually deal with the Hausdorff topological vector spaces in the sequel.

THEOREM 1. Let E be a topological vector space.

(a) For each $x_0 \in E$ and $\lambda \in \mathbb{K}$, $\lambda \neq 0$, the map $x \to \lambda x + x_0$ is a homeomorphism of E onto itself.

(b) For any subset A of E and any basis $\mathcal{B}$ of the neighbourhood filter at 0,

$$\bar{A} = \cap \{ A + V \; ; \; V \in \mathcal{B}\}$$

where $\bar{A}$ is the closure of A.

(c) If A is an open subset of E and B any subset of E, then A + B is an open subset of E.

(d) If A is a closed subset and B a compact subset of E, then A + B is a closed subset of E.

(e) If A is a circled subset of E, so is its closure $\overline{A}$.

THEOREM 2. Let E be a topological vector space. Then there exists a neighbourhood basis $\mathcal{B}$ of 0 in E such that

(N_1) each U in $\mathcal{B}$ is closed, circled and absorbing,

and (N_2) for each U in $\mathcal{B}$, there is a V in $\mathcal{B}$ with $V+V\subset U$.

Conversely, if E is a vector space and $\mathcal{B}$ is a filter basis satisfying (N_1) and (N_2), then there is a unique topology u on E which makes it a topological vector space and $\mathcal{B}$ is a neighbourhood basis at 0.

A topological vector space (E,u) is metrizable if there is a metric on E whose open balls form a basis. A topological vector space E is metrizable iff there is a countable neighbourhood basis at 0. These neighbourhoods can be so chosen as to satisfy (N_1) and (N_2) of Theorem 2.

A subset B of a topological vector space E is called (i) bounded if it is absorbed by every neighbourhood of 0 in E, (ii) totally bounded (precompact) it for each neighbourhood V of 0 in E, there is a finite subset B_0 in B such that $B\subset B_0+V$.

Every totally bounded subset of a topological vector space E is bounded.

A topological vector space E is said to be (i) complete if every Cauchy filter is convergent, (ii) quasi-complete if every closed and bounded subset of E is complete, and (iii) sequentially complete (or, semi-complete) if every Cauchy sequence in E converges.

Completeness $\Rightarrow$ quasi-completeness $\Rightarrow$ sequential completeness. (Cf. B [31])

If M is a closed subspace and N a finite dimensional subspace of a topological vector space E, then M+N is closed in E. If M is a closed subspace of finite codimension (that is, E/M is of finite dimension), then $E = M \oplus N$ for every algebraic complementary subspace N of M.

THEOREM 3. Every n-dimensional topological vector space E over the field $\mathbb{K}$ is topologically isomorphic to $\mathbb{K}^n$ with its natural topology. (Cf. B [31])

THEOREM 4. A topological vector space is finite dimensional iff it is locally compact. (Cf. B [31])

THEOREM 5. There is a one-to-one correspondence between the kernals of the continuous non-zero linear functionals on a topological vector space E and the closed hyperplanes in E. (Cf. B [31])

A topological vector space E is called a semiconvex space if it has a neighbourhood basis at 0 consisting of semiconvex sets. E is semiconvex iff there is a family $\{p_\alpha\}$ of (continuous) k_α-semi-norms $(0<k_\alpha\leq 1)$ such that the sets $\{x\in E \ ; \ p_\alpha(x)<1\}$ form a neighbourhood basis at 0.

A topological vector space E is called locally bounded if it has a bounded neighbourhood at 0. E is locally bounded iff its topology is given by a k-norm $(0<k\leq 1)$. Clearly a locally bounded space is semi-convex. A (Hausdorff) locally bounded space is metrizable. A product of an infinite family of locally bounded spaces $\neq \{0\}$ is not locally bounded (B [31], page 30).

A topological vector space E is called a locally convex space if it has a neighbourhood basis at 0 consisting of convex neighbourhoods. These neighbourhoods can be chosen to be closed and circled. Since there is a one-to-one correspondence between the sets of all closed, circled and convex subsets containing 0 as their interior point and the set of all continuous semi-norms in a topological vector space, it follows that a locally convex topology can also be defined by a set of continuous semi-norms on E.

Clearly a locally convex space is a locally semi-convex space.

1. A topology on a vector space, which is not compatible with the vector space structure.

The discrete topology on a vector space $E \neq \{0\}$ is not compatible with its vector space structure.

2. A topological vector space which is not a locally semi-convex space.

(i) Let E be the vector space of all measurable real functions on the closed interval $[0,1]$. Define the following metric on E:

$$d(f,g) = \int_0^1 \frac{|f(x) - g(x)|}{1+|f(x) - g(x)|}\, dx, \quad f,g \in E .$$

Then E, with the metric topology defined by d, is a topological vector space which is not a locally semi-convex space.

(ii) Let $p = \{p_i\}$ be a sequence of positive numbers, $0<p_i<1$ for all i, decreasing to 0. Let E be the set of all sequences $x = (x_i)$ of real numbers such that

$$\sum_{n=1}^{\infty} |x_i|^{sp_i} < \infty .$$

Definite the metric

$$d(x,y) = \sum_{i=1}^{\infty} |x_i - y_i|^{sp_i} .$$

Then E, with the metric topology defined by d, is a topological vector space which is not a locally semi-convex space.

3. A locally bounded (and hence a locally semi-convex) space which is not a locally convex space.

(i) Let $E = L^p[a,b]$, $a,b \in R$, $0<p<1$, be the space of all equivalence classes of measurable functions f(t) on $[a,b]$ with

$$\int_a^b |f(t)|^p dt < \infty .$$

Define

$$q_p(f) = (\int_a^b |f(t)|^p dt)^{1/p} .$$

Then E, equipped with the topology induced by the quasi-norm q_p, is a locally bounded space but not locally convex.

(ii) Similarly the space ℓ^p, $0<p<1$, of sequences $x=(x_n)$, $\sum_{n=1}^{\infty} |x_n|^p < \infty$, is a locally bounded space for the topology induced by q_p, where

$$q_p = (\sum_{n=1}^{\infty} |x_n|^p)^{1/p} .$$

But it is not locally convex.

(iii) Let H^p, $0<p<1$, denote the vector space of functions f(z) of the complex variable z, which are analytic in the interior of the unit circle and satisfy

$$\sup \{\int_0^{2\pi} |f(re^{i\theta})|^p \, d\theta \quad ; \quad 0 \leq r < 1\} < \infty.$$

Define

$$q_p(f) = \sup \{A_p(r;f) \quad ; \quad 0 \leq r < 1\}$$

where

$$A_p(r;f) = \left(\frac{1}{2\pi} \int_0^{2\pi} |f(re^{i\theta})|^p \, d\theta\right)^{1/p}.$$

Then (H^p, q_p) is a locally bounded space which is not a locally convex space.

4. A locally convex space which is not a locally bounded space.

$\mathbb{R}^{\mathbb{N}}$, equipped with the product topology, is a locally convex space which is not locally bounded.

5. A locally semi-convex space which is neither locally convex nor locally bounded.

Let ℓ^p, $0<p\leq 1$, be defined as in # 3(ii). Then the space

$$E = \ell^1 \times \ell^{\frac{1}{2}} \times \ell^{1/3} \times \ldots,$$

equipped with the product topology is a locally semi-convex space; but E is not locally bounded, because the product of an infinite family of topological vector spaces can never be locally bounded. Clearly E is not locally convex.

6. A topological vector space on which there exist no non-trivial continuous linear functionals.

Consider the topological vector space $L^p[a,b]$, $0<p<1$, as defined in # 3(i). We show that every continuous linear functional on L^p vanishes identically. Suppose $\phi \neq 0$ is a continuous linear functional on L^p. Then

$|\phi(f_0)| = 1$ for some $f_0 \varepsilon \underline{L}^p$. For $a < s < b$, we put

$$f_s^{(1)}(t) = \begin{cases} f_0(t) & \text{if } a \leq t \leq s \ , \\ 0 & \text{if } t > s \ . \end{cases}$$

Define

$$f_s^{(2)}(t) = f_0(t) - f_s^{(1)}(t).$$

Now, $q_p^p(f_s^{(1)})$, where

$$q_p(f_s^{(1)}) = (\int_a^s |f_0(t)|^p \, dt)^{1/p} \ ,$$

increases continuously from 0 to $q_p^p(f_0)$ so that there is an s_0 with

$$q_p^p \, (f_{s_0}^{(1)}) = q_p^p \, (f_{s_0}^{(2)}) = \tfrac{1}{2} \, q_p^p \, (f_0) \ .$$

Since $|\phi \, (f_0)| = 1$,

$$|\phi \, (f_{s_0}^{(i)})| \geq \tfrac{1}{2} \ , \ i = 1 \text{ or } 2.$$

Let

$$f_1(t) = 2 f_{s_0}^{(i)} \, (t) \text{ for this } i.$$

Then $|\phi \, (f_1)| \geq 1$ with

$$q_p(f_1) = 2^{p-1/p} \, q_p(f_0).$$

By induction, we define a sequence $\{f_n\}$ with $|\phi \, (f_n)| \geq 1$, and

$$q_p(f_n) = 2^{n(p-1)/p} \, q_p(f_0) \to 0$$

contradicting the continuity of ϕ .

7. A topological vector space such that no finite-dimensional subspace has a topological complement in it.

Let $E = \underline{L}^p[a,b]$, $0<p<1$, be as in # 3(i). Let H be a finite dimensional subspace of E and H_1 a one-dimensional subspace of H. Then, by Theorem 3, H_1 has a topological complement H_2 in H under the induced topology so that $H = H_1 \oplus H_2$. If G were a topological complement of H, then $H_2 \oplus G$ would be a topological complement of H_1 and so a closed hyperplane of E which is impossible by Theorem 5 and # 6.

8. Two closed subspaces of a topological vector space, whose sum is not closed.

Let E be a Hilbert space with an orthonormal basis $(e_n)_{n \geq 0}$.

Let $a_n = e_{2n}$ and

$$b_n = e_{2n} + \frac{1}{(n+1)} e_{2n+1}$$

for every $n \geq 0$; let A (respectively B) be the closed vector subspace of E generated by the a_n (respectively b_n). Show that:

(a) $A \cap B = \{0\}$; hence the sum A+B is direct (algebraically).

(b) The direct sum A+B is not a topological direct sum (consider in that subspace the sequence of points $b_n - a_n$ and consider projections of $A \times B$).

(c) The subspace A+B of E is dense but not closed in E (show that the point $\sum_0^\infty (b_n - a_n)$ does not belong to A+B).

(Cf. Page 123 of Dieudonne's "Foundations of Modern Analysis", Academic Press, Inc., New York, 1960).

9. A topological vector space in which the convex envelope of a precompact set is not precompact (not even bounded).

Consider $\ell^{\frac{1}{2}}$ with the usual norm. Let e_n be the element having 1 in the nth coordinate and zeros elsewhere. The set consisting of 0 and the points $n^{-\frac{1}{2}}e_n$ is compact whose convex envelope is unbounded, because the sequence of convex combinations

$$y_n = \frac{1}{n} \sum_{r=1}^{n} r^{-\frac{1}{2}} e_r$$

is unbounded.

10. A bounded linear map from a topological vector space to a topological vector space, which is not continuous.

Let $(E, ||\cdot||)$ be an infinite-dimensional Banach space. Then the identity map: $(E, \sigma(E,E')) \to (E, ||\cdot||)$ is bounded but not continuous.

CHAPTER 2

LOCALLY CONVEX SPACES

Introduction

We have already introduced the definition of a locally convex space in Chapter 1. Now we introduce some properties and results of locally convex spaces which will be used in the sequel.

A locally convex space E is called B-complete if every linear, continuous and almost open map of E onto any locally convex space is open. (A linear map f from E to F is almost open if, for each neighbourhood V of 0 in E, $\overline{f(V)}$ contains a neighbourhood of 0 in F). A complete and metrizable locally convex space is called a Fréchet space. A locally convex space whose topology is generated by a norm $||\cdot||$ is called a normed space. A complete normed space is called a Banach space. A vector space equipped with an inner product is called an inner product (or, pre-Hilbert) space. A complete inner product space is called a Hilbert space.

Hilbert space $\Longrightarrow$ Banach space $\Longrightarrow$ Fréchet space.

A vector space equipped with a semi-inner product is called a semi-inner product space. A semi-inner product space can be made into a normed space and a normed space can be made into a semi-inner product space. Every inner product space is a semi-inner product space. (Cf. P [24])

A vector space E is called a generalized inner product space (generalized semi-inner product space) if

(i) there is a subspace M of E which is an inner product space (respectively, a semi-inner product space) and (ii) there is a nonempty set $\mathcal{L}$ of linear operators on E such that (a) $\mathcal{L}E \subset M$ and (b) $Tx = 0$ for all T in $\mathcal{L}$ implies $x = 0$.

Every generalized inner product space is a generalized semi-inner product space. (Cf. p [72] , p [69])

Let $\mathcal{L}(E,F)$ denote the vector space of all continuous linear maps from a topological vector space E into a topological vector space F. Let $\mathfrak{S}$ be a class of subsets of E. We can define a topology in $\mathcal{L}(E,F)$ of uniform convergence over the sets of $\mathfrak{S}$ as follows: Let $\mathcal{B}$ be a neighbourhood basis at 0 in F. The family $\{M(S,V) \ ; \ S \in \mathfrak{S}, \ V \in \mathcal{B}\}$, where

$$M(S,V) = \{f \in \mathcal{L}(E,F) \ ; \ f(S) \subset V \},$$

is a neighbourhood basis at 0 in $\mathcal{L}(E,F)$ for a unique translation-invariant topology called the $\mathfrak{S}$-topology. $\mathcal{L}_{\mathfrak{S}}(E,F)$ is a topological vector space iff f(S) is bounded in F for each $S \in \mathfrak{S}$ and $f \in \mathcal{L}(E,F)$. In addition, if F is locally convex, so is $\mathcal{L}_{\mathfrak{S}}(E,F)$. If $\mathfrak{S}$ consists of bounded subsets of E such that $\bigcup_{\mathfrak{S}} S$ is total in E (that is, the linear envelope of $\bigcup_{\mathfrak{S}} S$ is dense in E) and F is a locally convex space, then $\mathcal{L}_{\mathfrak{S}}(E,F)$ is a Hausdorff locally convex space. If $\mathfrak{S}$ consists of all finite (compact, precompact, bounded) subsets of E, then the $\mathfrak{S}$-topology is called the topology of simple convergence (respectively, the topology of compact convergence, the topology of precompact convergence, the topology of uniform convergence on bounded sets).

Let E and F be a pair of vector spaces over $\mathbb{K}$. Let ϕ

be a bilinear functional on $E \times F$ satisfying the following separation axioms:

(D) $\phi(x,y) = 0$ for all $y \in F$ implies $x = 0$.

(D_1) $\phi(x,y) = 0$ for all $x \in E$ implies $y = 0$.

Then E and F are said to form a dual pair or dual system and we write $\langle E, F \rangle$. ϕ is called the canonical bilinear functional of the duality and is usually denoted by

$$(x,y) \to \langle x,y \rangle \quad .$$

Now, let E be a locally convex space and E' the (topological) dual of E, that is, the vector space of all continuous linear functionals on E. Clearly

$$E' \subset E^* \subset \mathbb{K}^E$$

where E^* is the algebraic dual of E, that is, the vector space of all linear functionals on E, and $\mathbb{K}^E$ is the product space.

The coarsest locally convex topology for which the map $x \to \langle x,x' \rangle$, for each $x' \in E'$, is continuous is called the weak topology on E; it is denoted by $\sigma(E,E')$ or w. Similarly we can define the so called week* topology $\sigma(E',E)$ or w^* on E'. $\sigma(E',E)$ is precisely the topology of simple convergence on E' which, in turn, is induced from the product topology of $\mathbb{K}^E$.

If $\mathfrak{S}$ consists of all bounded subsets of E, the $\mathfrak{S}$-topology on E' is denoted by $\beta(E',E)$ and is called the strong topology on E'. Similarly we can define the strong topology $\beta(E,E')$ on E.

Let A be a subset of a locally convex space. Then

$A^0 = \{x' \varepsilon E' \; ; \; \mathrm{Re}\langle x,x'\rangle \leq 1 \text{ for all } x \varepsilon E\}$

is called the polar of A. The bipolar A^{00} of A is defined by

$A^{00} = \{x \varepsilon E \; ; \; \mathrm{Re}\, \langle x,x'\rangle \leq 1 \text{ for all } x' \varepsilon A^0\}$.

THEOREM 1. Let A,B and $\{A_\alpha\}_{\alpha \varepsilon I}$ be subsets of a locally convex space E.

(a) $(\lambda A)^0 = \frac{1}{\lambda} A^0$, $\lambda \varepsilon \mathbb{K}$, $\lambda \neq 0$.

(b) If $A \subset B$, then $A^0 \supset B^0$ and $A^{00} \subset B^{00}$.

(c) $A \subset A^{00}$.

(d) $A^0 = A^{000}$.

(e) A^0 is convex, $\sigma(E',E)$ - closed and contains 0. If A is circled, so is A^0 .

(f) If A is a subspace of E, then A^0 is a $\sigma(E',E)$-closed subspace of E'. (Cf. B [31])

1. A locally convex space which is not metrizable.

Let I be an uncountable set of indices and $\mathbb{R}$ the real line. Then the product $\mathbb{R}^I$ is a locally convex space which is not metrizable, because the topological product of metrizable locally convex spaces is metrizable iff the product has finitely or countably many factors.

2. A metrizable topological vector space which is not locally convex.

The spaces considered in # 3(i), (ii), (iii) in Chapter 1 are metrizable topological vector spaces which are not locally convex.

3. A sequentially complete locally convex space which is not quasi-complete.

(i) Let w_d be the topological product of d, $d > \aleph_0$, copies of the field $\mathbb{K}$, where $\mathbb{K}$ is equipped with the natural topology. Let H be the subspace of w_d consisting of all vectors $x = (x_\alpha)$ with only countably many non-zero coordinates x_α. H is dense in w_d, and indeed every element of w_d is a closure point of H; H is, therefore, not quasi-complete, although it is sequentially complete.

(ii) $E = \ell^1 = \{x = (x_n), \sum_{n=} |x_n| < \infty\}$,

with its weak topology, is sequentially complete. It is not quasi-complete because if x" is an element of E" not generated by an element of E, then x" is the $\sigma(E",E')$-limit of a bounded directed family $\{x_i\}$ in E, and then the (weakly) closed convex envelope in E of $\{x_i\}$ is (weakly) bounded and (weakly) closed but not (weakly) complete.

4. A quasi-complete locally convex space which is not complete.

Let E be an infinite-dimensional Banach space. Then its dual E' is weakly quasi-complete but not weakly complete. (if E' were weakly complete, E would be finite-dimensional).

5. A complete locally convex space which is not B-complete.

Let E be an infinite-dimensional Bnach space and let E_w denote E equipped with the finest locally convex topology w. Let

$i : E_w \to E_u$

be the identity map. Then i is linear, one-to-one and onto. Since u w, i is continuous, and since E_u is barrelled, (see Chapter 3 for definition) i is almost open (because) a linear map from a locally convex space onto a barrelled space is almost open). We show that i is not open : If it were, any circled, convex and absorbing subset of E would be a u-neighbourhood of 0. In particular, any linear functional on E would be u-continuous. This is false: Take any infinite sequence (x_n) in E that is linearly independent and assume that $x_n = 1$ for each n. The sequence (x_n) may be extended into an algebraic base for E and a linear functional on E may be defined so as to take arbitrarily preassigned values at elements of the base. In particular, there is a linear functional f on E such that $f(x_n) = x_n$, $n = 1,2, \ldots$. Evidently this f is not u-continuous.

6. A complete locally convex space which is not metrizable.

(i) The locally convex space $\mathbb{R}^I$ of #1 is complete since each $\mathbb{R}$ is. But it is not metrizable as was shown there.

(ii) Consider the vector space $\phi = \mathbb{R}^{(\mathbb{N})}$ of all finite sequences equipped with the finest locally convex topology w. ϕ is the strict inductive limit (See Chapter 3 for definition) of a strictly increasing sequence of finite-dimensional Euclidean spaces $\mathbb{R}^n$ $(n \geq 1)$ which are Fréchet spaces. Since a locally convex space with the finest locally convex topology is complete, ϕ is complete.

But ϕ is not metrizable: It it were, it would then be a Fréchet space and hence of the second category. But ϕ is a countable union of non-dense subsets (because the identity map $\mathbb{R}^n \to \mathbb{R}^{n+1}$ is "into" for each n), and therefore of the first category which is a contradiction.

7. A normed space (and hence a metrizable locally convex space) which is not complete.

(i) The vect space $\phi = \mathbb{R}^{(\mathbb{N})}$ of all finite sequences with the supremum norm topology is a normed space which is not complete (See Chapter 3, # 16).

(ii) Let E = C(I) be the Banach space of continuous functions on the closed interval $I = [0,1]$, equipped with the supremum norm. Let F be the subspace of E consisting of functions f that vanish in a neighbourhood (depending on f) of t = 0, under the relative topology (See Chapter 6 for definition). Then F is a normed space which is not complete (See Chapter 5, # 23).

8. A locally convex space which contains a closed, circled and convex set with no extreme points.

(i) Let B denote the closed unit ball of the Banach space c_0 of sequences converging to 0. Then B does not have extreme points : If $x = (x_n)$ is in c_0 and

$$||x|| = \sup_n |x_n| = 1,$$

then we may replace some coordinate x_k, with $|x_k|<1$, alternately by $x_k + \varepsilon$ and $x_k - \varepsilon$ with ε sufficiently small. x, then, lies between the two points of B obtained in this way.

(ii) Let $L^1[a,b]$, $a,b\in\mathbb{R}$, be the vector space of all the equivalence classes of measurable functions f(t) on the closed interval $[a,b]$, with

$$\int_a^b |f(x)|\, dx < \infty .$$

$L^1[a,b]$ is a Banach space under the norm

$$||f|| = \int_a^b |f(x)|\, dx.$$

The unit ball B of $L^1[a,b]$ has no extreme points: Let

$$\int_a^b |f(x)|\, dx = 1, \quad f\in L^1[a,b] .$$

We determine c in such a way that

$$\int_a^b |f(x)|\, dx = \tfrac{1}{2} .$$

Let

$$f_1(x) = \begin{cases} 2f(x) & \text{if } x\in[a,c) \\ 0 & \text{if } x\in[c,b] \end{cases}$$

and

$$f_2(x) = \begin{cases} 0 & \text{if } x\in[a,c) \\ 2f(x) & \text{if } x\in[c,b] . \end{cases}$$

Then f is the middle point of the segment $[f_1, f_2]$ whose end points both belong to B.

9. A topological vector space which contains a compact convex set with no extreme points.

Cf. Roberts, J.W.: A compact convex set with no extreme points, Studia Mathematica, T. LX(1977), 255-264.

10. A weakly compact set in a locally convex space, whose weakly closed convex hull is not weakly compact.

Let $\phi = \mathbb{R}^{(\mathbb{N})}$ be the space of finite sequences with the usual supremum norm. Define

$$f_n(x) = x_n, \quad x = (x_n) \varepsilon \phi .$$

Then the set consisting of the points $2^{2n} f_n$ together with 0 is weakly compact, but is not strongly bounded. Thus, its w^*-closed convex hull is not w^*-compact.

11. A bounded sequence in a topological vector space, which is not convergent.

The sequence 0,1,0,1,0,1, ... in $\mathbb{R}$ is a bounded sequence which is not convergent.

CHAPTER 3

SPECIAL CLASSES OF LOCALLY CONVEX SPACES

Introduction

The concepts of Banach space, Hilbert space, inner (semi-inner) product space and generalized inner (semi-inner) product space are defined in the introduction to Chapter 2. In the present Chapter, we introduce some more concepts which are needed in the sequel. As we have remarked in the introduction to Chapter 1, we usually deal with Hausdorff topologies.

A topological vector space E is called a Baire space if it cannot be written as the union of an increasing sequence of nowhere dense sets. (A subset A of E is called nowhere dense if its closure $\bar{A}$ has empty interior). A locally convex space E is called a Baire-like space if it is not the union of an increasing sequence of nowhere dense, circled and convex sets. A complete metrizable locally convex space is called a Fréchet space. A Banach space is a Fréchet space; a Fréchet space is a Baire locally convex space; a Baire locally convex space is a Baire-like space.

A locally convex space E is called an unordered Baire-like space if it is not the union of a sequence of nowhere dense, circled and convex sets. An unordered Baire-like space is Baire-like.

Let $\{E_\alpha\}_{\alpha\varepsilon I}$ be a family of locally convex spaces, E a vector space and f_α a linear map from E_α into E, for each α. Let $E = \bigcup_\alpha f_\alpha(E_\alpha)$. The finest locally convex topology u for which each f_α is continuous is called the inductive limit of $\{E_\alpha\}_{\alpha \varepsilon I}$ with respect to the maps f_α. If $I = \mathbb{N}$, each f_n is the identity map and the inductive limit topology on E induces the same topology as that of E_n, then (E,u) is called the strict inductive limit of $\{E_n\}$.

The (strict) inductive limit of a properly increasing sequence of Banach (Fréchet) spaces is called the (strict) (LB)-space (respectively, (LF)-space).

A (Hausdorff) locally convex space E is t-polar if a subspace M of E' is weakly closed whenever $M \cap B^0$ is weakly closed for every barrel B in E.

A subset B of a locally convex space E is said to be bornivorous if it absorbs all bounded subsets of E. A closed, circled, convex and absorbing subset A of E is called a barrel. A locally convex space E is called a barrelled (quasi-barrelled) space if each barrel (bornivorous barrel) in E is a neighbourhood of 0. A Fréchet space is barrelled. A barrelled space is quasi-barrelled.

THEOREM 1. A sequentially complete quasibarrelled space is barrelled. (Cf. B [15])

THEOREM 2. Let E be a locally convex space. The following statements are equivalent:

(a) E is barrelled (quasi-barrelled).

(b) Each $\sigma(E',E)$-bounded ($\beta(E',E)$-bounded) subset of E' is equicontinuous.

(c) Each lower semi-continuous (bounded lower semi-continuous) semi-norm on E is continuous. (Cf. B [15])

A locally convex space E is called bornological if each circled, convex and bornivorous set in E is a neighbourhood of 0. A bornological space is quasi-barrelled. A metrizable locally convex space is bornological.

THEOREM 3. (a) A locally convex space E is bornological iff each bounded semi-norm on E is continuous.

(b) A locally convex space E is bornological iff each bounded linear map of E into any locally convex space F is continuous. (Cf. B[15] or B[31] or B[20])

PROPOSITION 1. The strong dual of a bornological space is complete. (Cf. B[15] or B[31] or B[20])

COROLLARY 1. Let E be a metrizable locally convex space. The following statements are equivalent:

(a) $(E', \beta(E',E))$ is bornological.

(b) $(E', \beta(E',E))$ is quasi-barrelled.

(c) $(E', \beta(E',E))$ is barrelled.

Let C(X) denote the locally convex space of continuous functions on a (Hausdorff) completely regular space X, equipped with compact-open topology.

THEOREM 4. C(X) is barrelled iff for every noncompact closed subset A of X, there exists some f in C(X) which is unbounded on A. (Cf. P [66])

THEOREM 5. C(X) is bornological iff X is a Q-space (real-compact space). (Cf. P [85]).

REMARK. For the definition of a Q-space (real-compact space), see B[18] or P[85] .

A locally convex space E is semi-bornological if every bounded linear functional on E is continuous. E is sequentially bornological (in short, S-bornological) if each convex and bornivorous set B in E is a sequential neighbourhood of 0 (that is, every sequence converging to 0 belongs to B eventually).

THEOREM 6. Let E be a locally convex space. The following statements are equivalent:

(a) E is S-bornological.

(b) Every circled, convex and bornivorous set in E is a sequential neighbourhood of 0.

(c) Every bounded semi-norm on E is sequentially continuous. (Cf. P[88])

A locally convex space E is said to be convex-sequential (in short, C-sequential) if every convex sequentially open subset of E is open.

THEOREM 7. Let E be a locally convex space. The following statements are equivalent:

(a) E is C-sequential.

(b) Every convex, circled and sequentially open subset of E is open.

(c) Every sequentially continuous semi-norm on E is continuous.

(d) Every convex, circled and sequential neighbourhood of 0 in E is a neighbourhood of 0.

(e) Every convex sequential neighbourhood of 0 in E is a neighbourhood of 0. (Cf. P [88])

THEOREM 8. A locally convex space is bornological iff it is both C-sequential and S-bornological. (Cf. P [88]).

The $\mathfrak{S}$-topology on a locally convex space (E,u) is called the Mackey topology if $\mathfrak{S}$ is the family of all circled, convex and $\sigma(E',E)$-compact subsets of E', and is denoted by $\tau(E,E')$. If $u = \tau(E,E')$, (E,u) is called a Mackey space.

THEOREM 9. A locally convex space E is a Mackey space iff each convex and $\sigma(E',E)$-relatively compact subset of E' is equicontinuous. (Cf. B [20])

A quasi-barrelled space is a Mackey space.

A locally convex space E is called quasi-M-barrelled if, in E', every circled, convex and $\beta(E',E)$-bounded set is $\sigma(E',E)$-relatively compact (equivalently, if the strong bidual induces the Mackey topology on E). A quasi-barrelled space is quasi-M-barrelled. E is said to have property (S) if $(E', \sigma(E',E)$ is sequentially complete and property (C) if every $\sigma(E',E)$-bounded subset of E' is $\sigma(E',E)$-relatively countably compact. If $(E'_\beta)' = E$ (algebraically), where β is the strong topology $\beta(E',E)$, then E is called a semi-reflexive space. If $(E'_\beta)'_\beta = E$ (topologically), then E is called a reflexive space. A reflexive space is clearly semi-reflexive.

THEOREM 10. (a) A locally convex space E is semi-reflexive iff each $\sigma(E,E')$-closed and bounded set is $\sigma(E,E')$-compact.

(b) A locally convex space E is reflexive iff E is

barrelled and each $\sigma(E,E')$-bounded set in E is $\sigma(E,E')$-relatively compact. (Cf. B [31])

A locally convex space E is a Montel space if E is barrelled and each bounded set in E is relatively compact. A Montel space is reflexive. E is called a Schwartz space if for every closed, circled and convex neighbourhood U of 0 in E, there exists a neighbourhood V of 0 such that for every $\alpha>0$, the set V can be covered by finitely many translates of αU.

THEOREM 11. A locally convex space E is a Schwartz space iff the following condition is satisfied: (Cf. B[13])

(S) Every bounded subset of E is totally bounded and for every closed, circled and convex neighbourhood U of 0, there is a neighbourhood V of 0 such that for every $\alpha>0$, we can find a bounded subset A of E such that $V \subset \alpha U + A$.

A locally convex space E is called a distinguished space if each $\sigma(E'',E')$-bounded subset of its strong bidual E'' is contained in the $\sigma(E'', E')$ - closure of some bounded subset of E (equivalently, if for each $\sigma(E'',E')$-bounded subset B of E'', there is a bounded subset A of E such that $B \subset A^{00}$, where A^{00} is the polar of A^{0} with respect to the dual pair $\langle E',E''\rangle$). Normed and semi-reflexive spaces are distinguished.

THEOREM 12. A locally convex space E is distinguished iff $(E', \beta(E',E))$ is barrelled. (Cf. B[15] or B [20] or B [31]).

COROLLARY 2. A metrizable locally convex space is distinguished iff $(E', \beta(E',E))$ is bornological.

A locally convex space E is countably barrelled (countably quasi-barrelled) if each $\sigma(E',E)$ - bounded ($\beta(E',E)$ - bounded) subset of E', which is the countable union of equicontinuous subsets of E', is itself equicontinuous (equivalently, if each barrel (bornivorous barrel) which is the countable intersection of circled, convex and closed neighbourhoods of 0 is a neighbourhood of 0). A barrelled (quasi-barrelled) space is countably barrelled (countably quasi-barrelled). A countably barrelled space is countably quasi-barrelled. A countably quasibarrelled space which has fundamental sequence of bounded sets is called a (DF)-space. A normed space is a (DF)-space. The strong dual of a metrizable locally convex space is a (DF)-space. A locally convex space E is σ-barrelled (σ-quasi-barrelled) if each $\sigma(E',E)$-bounded ($\beta(E',E)$-bounded) sequence in E' is equicontinuous. A countably barrelled (countably quasi-barrelled) space is σ-barrelled (σ-quasi-barrelled). A σ-barrelled space is σ-quasi-barrelled. E is sequentially barrelled if each $\sigma(E', E)$-convergent sequence in E' is equicontinuous. A σ-barrelled space is sequentially barrelled. E is an H-space if its strong dual is countably barrelled. Distinguished spaces and metrizable locally convex spaces are H-spaces. E is k-barrelled (k-quasi-barrelled) if the intersection of a sequence $\{V_n\}$ of circled, convex and closed neighbourhoods of 0 is a barrel (bornivorous barrel) implies that $\bigcap_{n=1}^{\infty} k^n V_n$ is a neighbourhood of 0. A k-barrelled space is clearly k-quasi-barrelled.

THEOREM 13. Let X be a completely regular Hausdorff

space. C(X) is countably barrelled iff every C(X)-pseudo-compact subset of X which is the closure of a countable union of compact sets is actually compact. (Cf. B[15])

1. An inner product (a pre-Hilbert) space which is not a Hilbert space.

Consider the space C[a,b], a,b$\in\mathbb{R}$, of continuous functions on the closed interval [a,b]. Define

$$(f,g) = \int_a^b f(t) . \overline{g(t)}\, dt.$$

Then C[a,b] becomes an inner product space which is not complete.

2. A generalized inner product space which is not an inner product space.

Let E = C ($\mathbb{R}$) be the space of all continuous real-valued functions on $\mathbb{R}$. Let M be the subspace consisting of all square integrable functions in E. We define the inner product in M as follows:

$$(x,y) = \int_{\mathbb{R}} x(t)\, y(t)\, dt.$$

We denote by τ the family of all projections P(I) defined by

$$(P(I)(x))\ (t) = f_I(t)\ x(t)$$

where, for every compact interval I, we choose an Uhryson function f_I with compact support which is 1 on I. Then E is a generalized inner product space which is not an inner product space.

3. A semi-inner product space which is not an inner product space.

Let $L^p(\mathbb{R})$, $2\leq p<\infty$, be the vector space of all the equivalence classes of measurable functions $f(t)$ on $\mathbb{R}$, with

$$\int_{\mathbb{R}} |f(t)|^p dt<\infty .$$

Define

$$[g,f] = \frac{1}{||f||_p^{p-2}} \int_{\mathbb{R}} g|f|^{p-1} \operatorname{sgn}(f)\, dt$$

where

$$||f||_p = \left(\int_{\mathbb{R}}|f(t)|^p\, dt\right)^{1/p} .$$

Then $[g,f]$ is a semi-inner product. But it is not an inner product.

4. A generalized semi-inner product space which is neither a semi-inner product space nor a generalized inner product space.

Let E be the space of all measurable functions on $\mathbb{R}$. Let $M = L^p(\mathbb{R})$, $2\leq p<\infty$, be as defined in # 3, with the semi-inner product in M as defined there. Let τ be the family of operators $E^{p-1}(I)$ such that for all $x,y\in E$ and for any scalars α and β,

$$E^{p-1}(I)\ (\alpha x + \beta y)\ (t)$$

$$= \chi_I(t)\ (|x(t)|^{p-2} + |y(t)|^{p-2})\ (\alpha x(t) + \beta y(t))$$

where I is the finite non-degenerate interval and $\chi_I(t)$ is the characteristic function of I. Then it is easy to check that E is a generalized semi-inner product space which is neither a generalized inner product space nor a semi-inner product space.

5. A Banach space which is not a Hilbert space.

(i) The space $L^p[0,1]$, $1 \leq p \leq \infty$, $p \neq 2$, of equivalence classes of pth power summable functions on the closed interval $[0,1]$ with the norm

$$||f|| = (\int_0^1 |f(t)|^p \, dt)^{1/p}, \quad f \in L^p\,[0,1]$$

is a Banach space which is not a Hilbert space.

(ii) The spaces c and c_0 of all convergent sequences and all sequences converging to 0 respectively are Banach spaces which are not Hilbert spaces.

6. A Banach space which is not separable.

The space ℓ^∞ of bounded sequences with the supremum norm topology is a Banach space. Two vectors in ℓ^∞ whose coordinates are equal to +1 and -1 are always distance 2 apart. Since there is a continuum of these, the set of these vectors is not separable, and so ℓ^∞ is not separable. (If a normed space is separable, so is every subset of it).

7. A Banach space which is not reflexive.

(i) The space c_0 of all sequence converging to 0, equipped with the supremum norm topology, is a Banach space which is not reflexive. For, the dual of c_0 is

$$\ell^1 = \{x = (x_n) \; ; \; \sum_{n=1}^{\infty} |x_n| < \infty \}$$

with the norm

$$||x|| = \Sigma|x_n| \; ,$$

and the dual of ℓ^1 is the Banach space ℓ^∞ as defined in # 6. Thus, ℓ^∞ is the bidual of c_0 , which is clearly larger than c_0 .

(ii) From the fact that c_0 is not reflexive, it follows that ℓ^1 and ℓ^∞ are not reflexive.

8. A Fréchet space which is not a Banach space.

Consider $E = \prod_{n=1}^{\infty} \mathbb{R}_n$ with the product topology, where each $\mathbb{R}_n$ is a copy of the real line equipped with the usual topology. Then E is a Fréchet space, but not a Banach space because a product of Banach spaces is a Banach space iff it is a product of a finite number of Banach spaces.

9. A t-polar space which is not B-complete.

Let $E = \phi$, the space of finite sequences, equipped with the supremum norm topology. Then $E' = \ell^1$. $(E', \sigma(E',E))$ is t-polar: For, the weakly bounded subsets of $E'' = E$ are exactly those which are bounded in the norm topology of E. Furthermore, if L is a subspace of E, whose intersection with the unit sphere is weakly closed (and hence closed), then L itself is closed, and therefore also closed in the weak topology $\sigma(E,E')$. $(E',\sigma(E',E))$ is not B-complete: The equicontinuous subsets of $E = E''$ are all of finite-dimension, and hence any subspace L of E satisfies the condition that $L \cap U^0$ is weakly closed for all neighbourhoods U of 0 in E'.

10. A barrelled space which is not complete.

Let E be a separable infinite-dimensional Banach space (eg. $E = \ell^1$). Then E contains a dense subspace M of countable infinite codimension, which is a Baire space. M is a barrelled space which is not complete.

11. A barrelled space which is not metrizable.

Let $E = \phi$, the space of finite sequences, equipped with the finest locally convex topology ν. Then (E,ν) is a barrelled space. Also it is complete (B[14], page 67). We show that it is not metrizable: It is easy to see that E is the strict inductive limit of a strictly increasing sequence of finite-dimensional Euclidean spaces $\mathbb{R}^n$ which are Fréchet spaces. Now suppose E is metrizable, then E is a Fréchet space and hence of the second category. But, since the identity map $\mathbb{R}^n \to \mathbb{R}^{n+1}$ is "into" for each n, E is a countable union of non-dense subsets and hence of the first category which is a contradiction.

12. A Baire-like space which is not an unordered Baire-like space.

Let w be the space of all real sequences with the product topology and let $E = \{(a_p)\varepsilon w;\ a_p = 0$ whenever $p\notin(n_k)$, for some (n_k) with $\lim (n_k/k) = 0\}$. Then E is a dense vector subspace of w and that every countable $\sigma(E',E)$-bounded subset of E' is equicontinuous, so that E is σ-barrelled. Since E has its weak topology, E is Baire-like (Cf. Todd, A.R. and Saxon, S.A. : A product of locally convex spaces, Math. Ann., 206(1973), 23-34). But E is clearly the union of the closed, proper (and hence nowhere dense) subspaces $F_n = \{(a_p)\varepsilon E;\ a_n = 0\}$, and thus E is not unordered Baire-like.

13. A Baire-like space which is not a Baire space.

Let $\{e_n\}$ be the unit vectors in ℓ^1. If the scalar sequence $\{b_n\}\varepsilon\ell^1$ has infinitely many non-zero entries, then the span E in ℓ^1 of $\{e_n\} \cup \{ \sum_{i=1}^{\infty} b_i e_{n_i} \}$, where n_i ranges over all subsequences of the sequence $\{1,2,\ldots\}$, is dense in ℓ^1 and so its dual is $m = \ell^\infty$. If we assume that some pointwise bounded subset B of E' = m is not norm-bounded, then we may use a "sliding hump" argument (Cf. P[77]) to obtain a sequence $\{h_k\} \subset B$ and a subsequence $\{n_i\}$ of $\{1,2,\ldots\}$ such that

$$| h_k(\sum_{i=1}^{\infty} b_i e_{n_i}) | \to \infty$$

contradicting the fact that B is $\sigma(E',E)$-bounded. Thus, E is barrelled. Let f_k be that unique member of E' such that

$$f_k(e_j) = \delta_{kj} , k,j = 1,2,\ldots .$$

We choose the sequence $\{b_i\}\varepsilon\ell^1$ so that the number of zero entries between the nth and $(n + 1)^{th}$ non-zero entries is a strictly increasing function of n (n = 1,2,...). Then, we readily see that every finite linear combination of the sort defining E has a zero entry, and thus

$$E = \bigcup_{k=1}^{\infty} f_k^{-1} (\{0\}).$$

But it is clear that each $f_k^{-1}(\{0\})$ is a closed proper 1-condimensional subspace in E, and hence is nowhere dense in E. So, E is not a Baire space. But it is a Baire-like space, being a normed barrelled space.

14. A barrelled bornological space which is not the inductive limit of Banach spaces.

Cf. Valdivia, M. : A class of bornological barrelled

spaces which are not ultrabornological, Math. Ann., 194 (1971), 43-51.

REMARK. The inductive limit of Banach spaces is also called ultrabornological. But we use this name for different concept in Chapter 4.

15. A bornological space which is not metrizable.

Let $E = \prod_{t\in[0,1]} \mathbb{R}_t$, where $\mathbb{R}_t$ is a copy of the real line $\mathbb{R}$. Since E is an uncountable product of metrizable spaces, it is not metrizable (A product of metrizable locally convex spaces is metrizable iff it is a countable product). However, E is bornological, since the cardinality $2^{\aleph_0}$ of $[0,1]$ is smaller than the smallest strongly inaccessible cardinal. (A cardinal d_0 is strongly inaccessible if (i) $d_0 > \aleph_0$, (ii) $\Sigma\{d_\alpha ; \alpha \in A\} < d_0$ whenever cardinal of $A < d_0$ and $d_\alpha < d_0$ for all $\alpha \in A$, (iii) $d < d_0$ implies $2^d < d_0$. See B[20], page 392).

16. A bornological space which is not barrelled.

The space ϕ of finite sequences equipped with the supremum norm topology is a normed space and hence bornological. But it is not barrelled. For, the sequence $\{f_n\}$ of continuous linear functionals, defined by

$$f_n(x) = x_n , \quad x = (x_n) \in \phi ,$$

is weakly bounded but not equicontinuous.

17. A barrelled space which is not bornological.

Let $W(\omega_2)$ be the space of all ordinals less than the initial ordinal ω_2 of the fourth class with the interval topology and let X be the subspace of $W(\omega_2)$ whose elements are not ω_0-limits. Then the space X is $\aleph_1$-additive and is not a Q-space, since it has no complete structures. Furthermore, by the normality and the $\aleph_1$-additivity of X, any closed and bounded subset of X is finite and hence satisfies the condition of Theorem 4. Thus C(X) is a barrelled space which is not bornological.

18. A quasi-barrelled space which is neither barrelled nor bornological.

Let E be a barrelled space which is not bornological and F a bornological space which is not barrelled. Then the product E×F is clearly a quasi-barrelled space. However, E×F is neither barrelled nor bornological: Let B be a bornivorous convex set in E, which is not a neighbourhood of 0 in E. Then B×F is a bornivorous convex set in E×F, which is not a neighbourhood of 0 in E×F. This shows that E×F is not bornological. Next, let B_1 be a barrel in F, which is not a neighbourhood of 0 in F. Then $E \times B_1$ is a barrel in E×F, which is not a neighbourhood of 0 in E×F. This shows that E×F is not barrelled.

19. A quasi-M-barrelled space which is not quasi-barrelled.

The requirement of being quasi-M-barrelled depends only on the dual system and so, weakening the topology of quasi-M-barrelled space without affecting the dual, would still leave it quasi-M-barrelled. Let E be an infinite-dimensional Banach space. Then $(E, \sigma(E,E'))$ is a quasi-M-barrelled space which is not quasi-barrelled.

20. A semi-bornological space which is not an S-bornological space (and hence not a bornological space).

Consider $\ell^2 = \{x = (x_n);\ \sum_{n=1}^{\infty} |x_n|^2 < \infty\}$.

Then $(\ell^2, \sigma(\ell^2, \ell^2))$ is a semi-bornological space which is not S-bornological : Let u be the norm topology on ℓ^2 . Since the space (ℓ^2, u) is normable, the space $(\ell^2, \sigma(\ell^2, \ell^2))$ is semi-bornological (B[37], page 190). We show that $(\ell^2, \sigma(\ell^2, \ell^2))$ is not S-bornological. For this, we must find a convex, circled and $\sigma(\ell^2, \ell^2)$-bornivorous set in ℓ^2 which is not a $\sigma(\ell^2, \ell^2)$-sequential neighbourhood of 0 in ℓ^2 . Consider the unit ball

$$B = \{x \varepsilon \ell^2 \ ; \ ||x|| < 1\} .$$

B is a convex, circled and u-neighbourhood of 0 in ℓ^2 . Consequently B is a convex, circled and u-bornivorous set. Since the $\sigma(\ell^2, \ell^2)$-bounded subsets of ℓ^2 are the same as the u-bounded subsets of ℓ^2 , B is a convex, circled and $\sigma(\ell^2, \ell^2)$-bornivorous set in ℓ^2 . We show that B is not a $\sigma(\ell^2, \ell^2)$-sequential neighbourhood of 0 in ℓ^2 . Consider the sequence (e_n) of unit vectors in ℓ^2 . e_n is the sequence in $\mathbb{K}$ with 1 in the nth place and 0 in all other places. The sequence (e_n) is weak convergent to the zero sequence in $\mathbb{K}$, that is, (e_n) is $\sigma(\ell^2, \ell^2)$-convergent to 0 in ℓ^2 . Since each element e_n of this sequence has norm 1 and is not in B, (e_n) is not ultimately in B. Clearly, B is not a $\sigma(\ell^2, \ell^2)$-sequential neighbourhood of 0 in ℓ^2.

21. An S-bornological space which is not C-sequential (ant hence not a bornological space).

The space $(\ell^1, \sigma(\ell^1, \ell^\infty))$ is S-bornological: We show that $(\ell^1, (\ell^1, \ell^\infty))$ is a braked space .

(E is a braked space if given any sequence $\{x_n\}$ in E, which converges to 0, there exists a sequence of positive real numbers $\{\lambda_n\}$ such that $\lambda_n \to +\infty$ and the sequence $\{\lambda_n x_n\}$ in E converges to 0). But then, it is S-bornological, because a braked space is S-bornological (P[88], page 279). Let u be the norm topology on ℓ^1. Let $\{x^{(n)}\}$ be a sequence in ℓ^1 converging to 0 in the weak topology $\sigma(\ell^1, \ell^\infty)$. Then $\{x^{(n)}\}$ is u-convergent to 0, because weak convergence and norm convergence of sequences in ℓ^1 are the same. Since (ℓ^1, u) is normable, it is braked. Consequently there exists a sequence of positive real numbers $\{\lambda_n\}$ such that $\lambda_n \to +\infty$ and the sequence $\{\lambda_n x^{(n)}\}$ is u-convergent to 0. Clearly $\{\lambda_n x^{(n)}\}$ is $\sigma(\ell^1, \ell^\infty)$-convergent to 0. Hence $(\ell^1, \sigma(\ell^1, \ell^\infty))$ is a braked space and hence an S-bornological space. $(\ell^1, \sigma(\ell^1, \ell^\infty))$ is not C-sequential because the unit ball

$$B = \{x \varepsilon \ell^1 \ ; \ ||x|| < 1\}$$

is a convex, circled and $\sigma(\ell^1, \ell^\infty)$-sequentially open subset of ℓ^1 which is not $\sigma(\ell^1, \ell^\infty)$-open.

22. A C-sequential locally convex space which is not S-bornological (and hence not bornological).

Consider the space $(\ell^2, \sigma(\ell^2, \ell^2)_{cs})$ which is the C-sequential locally convex space generated by the locally convex space $(\ell^2, \sigma(\ell^2, \ell^2))$. We show that $(\ell^2, \sigma(\ell^2, \ell^2)_{cs})$ is not S-bornological. For this we must find a convex, circled and $\sigma(\ell^2, \ell^2)_{cs}$ -bornivorous set in ℓ^2 which is not a $\sigma(\ell^2, \ell^2)_{cs}$-sequential neighbourhood of 0 in ℓ^2. As shown in # 20, the unit ball

$B = \{x \in \ell^2 \quad ; \quad ||x||<1\}$

is a convex, circled and $\sigma(\ell^2, \ell^2)$-bornivorous set in ℓ^2. Since $(\ell^2, \sigma(\ell^2, \ell^2))$ and $(\ell^2, \sigma(\ell^2, \ell^2)_{cs})$ have the same convergent sequences, they have the same bounded sets. This is because of the sequential characterization of bounded sets. A set A in a topological vector space is bounded iff given any sequence $\{x^{(n)}\}$ in A and given any sequence $\{\lambda_n\}$ of positive real numbers such that $\lim_{n\to+\infty} \lambda_n = 0$, we have $\lim_{n\to+\infty} \lambda_n x_n = 0$. Consequently B is a convex, circled and $\sigma(\ell^2, \ell^2)_{cs}$-bornivorous set in ℓ^2. Of course, B is not a $\sigma(\ell^2, \ell^2)_{cs}$-sequential neighbourhood of 0 in ℓ^2. The sequence $\{e_n\}$ of unit vectors in ℓ^2 is $\sigma(\ell^2,\ell^2)_{cs}$-convergent to 0 in ℓ^2, since it is $\sigma(\ell^2, \ell^2)$-convergent to 0 in ℓ^2. However, every term e_n of this sequence has norm 1 and is, therefore, not in B.

23. A Mackey space which is not quasi-barrelled.

Cf. # 26 , and # 37.

24. A Mackey space which does not have property (S).

The locally convex space $(\ell^1, \tau(\ell^1, c_0))$, where $\tau(\ell^1, c_0)$ is the Mackey topology, is a Mackey space which does not have property (S), because $(c_0, \sigma(c_0, \ell^1))$ is not sequentially complete.

25. A Mackey space with property (S) but without property (C).

The Mackey space $(\ell^\infty, \tau(\ell^\infty, \ell^1))$ has property (S) (Cf. B[5], Page 374). To show that it does not have property (C), we proceed as follows: Let

$$B = \{e_n \ ; \ n = 1,2,\ldots\}$$

be the canonical Schauder basis of ℓ^1, where e_n is the sequence of all zeros except the nth coordinate which is 1. B is $\sigma(\ell^1, \ell^\infty)$-bounded, but has no $\sigma(\ell^1, \ell^\infty)$-accumulation point in ℓ^1. For, suppose $y \varepsilon \ell^1$. There exists a sequence $\Phi \varepsilon \ell^\infty$ of the form

$$\Phi = (0, \ldots, 0,2,2, \ldots)$$

such that $| < \Phi, y > | < 1$. Thus,

$$< \Phi, e_n - y > = < \Phi, e_n > - < \Phi, y > > 1$$

holds true for n big enough that $< \Phi, e_n > = 2$. Then

$$e_n - y \varepsilon \{\Phi\}^0$$

for atmost finitely many $n \varepsilon N$. Therefore y is not a $\sigma(\ell^1, \ell^\infty)$-accumulation point of B.

26. A semi-reflexive space which is not reflexive.

Let E be a non-reflexive Fréchet space (for example, $E = c_0$). Then the Mackey space $(E', \tau(E',E))$ is a semi-reflexive space which is not reflexive, because its strong dual E is not reflexive, (The strong dual of a reflexive space is reflexive).

27. A barrelled space which is not a Montel space.

An infinite-dimensional Banach space is clearly a barrelled space. However, it is not a Montel space, because a normed Montel space is locally compact and hence finite-dimensional.

28. A reflexive space which is not a Montel space.

Any infinite-dimensional reflexive Banach space is an example of a reflexive space which is not a Montel space. (For example, consider the space ℓ^p, $1< p<\infty$).

29. A Fréchet space which is not a Schwartz space.

Let E be any infinite-dimensional Banach space. As shown in # 27, E is not a Montel space. Clearly E is a Fréchet space. Since a Fréchet Schwartz space is a Montel space, it is clear that E is not a Schwartz space.

30. A Schwartz space which is not a Montel space.

Let E be any infinite-dimensional vector space with algebraic dual $E^* = F$. We show that $(E, \sigma(E,F))$ is a Schwartz space. For this, we show that it satisfies the condition (S) of Theorem 11: Every bounded subset of E is relatively compact in $(F^*, \sigma(F^*,F))$ and thus precompact in E. Let

$$U = \{x \; ; \; |\langle x, y_k\rangle| \leq \varepsilon, \quad 1 \leq k \leq n\}$$

be a neighbourhood of 0, where we may suppose that the vectors y_k are linearly independent. Let M be the subspace of F generated by the y_k $(1\leq k\leq n)$, N an algebraic supplement of M and $\{z_\ell\}_{\ell\varepsilon I}$ a basis of N. We choose V = U and

$$A = \{x \; ; \; |\langle x, y_k\rangle| \leq \varepsilon, \quad 1 \leq k \leq n,$$
$$|\langle x, z_\ell\rangle| \leq \varepsilon, \quad \ell \varepsilon I\}$$

for the set A which appears in the condition (S) of Theorem 11, indpendently of α. In the first place, A is bounded because, if

$$y = \sum_{k=1}^{n} \eta_k y_k + \sum_{\ell \varepsilon I} \zeta_\ell z_\ell \varepsilon F,$$

We have

$$|\langle x,y \rangle| \leq \varepsilon\left(\sum_{k=1}^{n} |\eta_k| + \sum_{\ell \varepsilon I} |\zeta_\ell|\right).$$

By an elementary algebraic consideration, there exists a basis $\{x_k\ ;\ 1 \leq k \leq n\}$ in N for which

$$\langle x_k, y_\ell \rangle = 0 \text{ if } k \neq \ell$$

and

$$\langle x_k, y_k \rangle = 1.$$

For each $x \varepsilon U$, set

$$u = \sum_{k=1}^{n} \langle x, y_k \rangle x_k.$$

Then, $u \varepsilon A$, since

$$|\langle u, y_k \rangle| = |\langle x, y_k \rangle|$$

and

$$\langle u, z_\ell \rangle = 0.$$

Finally

$$x - u \varepsilon \alpha U$$

since

$$\langle x - u, y_k \rangle = \langle x, y_k \rangle - \langle x, y_k \rangle = 0.$$

Thus,

$$U \subset \alpha U + A.$$

Hence $(E, \sigma(E,F))$ is a Schwartz space by Theorem 11. But it is not barrelled (See # 47).

31. A Montel space which is not separable.

The space ϕ_d, $d > \aleph_0$, of all sequences with d non-zero coordinates, equipped with the finest locally convex topology, is a Montel space which is not separable.

32. A Montel space (and hence a reflexive locally convex space) which is not complete.

Consider the following notations:

$\{d_n\}$ = the sequence of cardinal numbers such that

$$d_0 = 2^{\aleph_0} \quad \text{and} \quad d_n = 2^{d_{n-1}},$$

$$\{\omega_n\} = \prod_{\alpha \varepsilon A_n} \mathbb{R}_\alpha \ ,$$

$$\{\phi_n\} = \bigoplus_{\alpha \varepsilon A_n} \mathbb{R}_\alpha \ ,$$

where each $\mathbb{R}_n$ is the real line and A_n is an index set of power d_n. $\omega_{n,0} = \{(x_\alpha) \varepsilon \omega_n \ ; \ x_\alpha = 0$ except for countable $\alpha\}$,

$\{e^n_\alpha\}$ = a Hamel base of $\omega_{n,0}$.

$\{\bar{e}^n_\alpha\}$ = the base of ϕ_n, that is, $\bar{e}^n_\alpha = (x_\beta)$ such that

$x_\alpha = 1$ and $x_\beta = 0$ for $\beta \neq \alpha$.

$\{\tilde{e}^n_\alpha\}$ = a Hamel base of ω_{n-1} such that for any $\chi \neq 0$, $\chi \varepsilon \phi_{n-1}$, the set $\{\tilde{e}^n_\alpha \ ; \ \langle \tilde{e}^n_\alpha , \chi \rangle \neq 0\}$ is uncountable.

S_n = the linear operator on $\omega_{n,0}$ to ϕ_n such that $S_n e^n_\alpha = \bar{e}^n_\alpha$.

T_n = the linear operator on ϕ_n to ω_{n-1} such that $T_n \bar{e}^n_\alpha = \tilde{e}^n_\alpha$.

$\varepsilon^n_\alpha = {}^tT_n \bar{e}^{n-1}_\alpha$.

E_n = the vector subspace of $\omega_n \times \omega_{n+1}$ generated by $\{(e^n_\alpha , \varepsilon^{n+1}_\alpha)\}$.

$E = \prod_n E_n$.

$F = \oplus_n \phi_n$.

$(E, \sigma(E,F))$ is a Montel space which is not complete (Cf. P[50] for details).

33. A distinguished space which is not semi-reflexive.

Let E be a non-reflexive Banach space (for example, $E=c_0$). Since E is a normed space, it is distinguished. But E is not semi-reflexive, because a locally convex space is reflexive iff it is semi-reflexive and quasi-barrelled.

34. A Fréchet space which is not distinguished.

Let E be the vector space of all numerical double sequences $x = (x_{ij})$ such that for each $n \in N$,

$$p_n(x) = \sum_{i,j} |a_{ij}^{(n)} x_{ij}| < +\infty ,$$

where

$$a_{ij}^{(n)} = \begin{cases} j & \text{for } i \leq n \\ 1 & \text{for } i > n . \end{cases}$$

The semi-norms $\{p_n\}$ generate a locally convex topology under which E is a Fréchet space. The dual E' can be identified with the space of double sequences $u = (u_{ij})$ such that $|u_{ij}| \leq c\, a_{ij}^{(n)}$ for all i,j and suitable $c>0$, $n \in N$. Let $B_n = U_n^o$ where $U_n = \{x \,;\, p_n(x) \leq 1\}$, and W the convex circled hull of $\bigcup_n 2^{-n} B_n$. Then W is bornivorous in E' and W contains no element $u \in E'$ with the property that, for each i, there exists j with $|u_{ij}| \geq 2$. We show that W can contain no strong neighbourhood of 0 : For each strong neighbourhood B^o in E', B a bounded subset of E, there exists a sequence $\rho = (\rho_n)$ of strictly positive numbers such that $\Gamma_n \rho_n B_n \subset B^o$. Now define elements $u^{(n)} \varepsilon E'$ so that

$$u_{ij}^{(n)} = 0 \quad \text{for } (i, j) \neq (n, k_n)$$

and

$$u_{n,k_n}^{(n)} = 1$$

where k_n is chosen so that

$$2^{n+1} u^{(n)} \varepsilon \rho_n B_n .$$

For each given ρ , the sequence with general term $s_N = 2 \sum_1^N u^{(n)}$ is a weak Cauchy sequence in E' and hence convergent to $s \varepsilon E'$. Now, $s_N \varepsilon B^o$ for all $N \varepsilon \mathbb{N}$. Hence $s \varepsilon B^o$ but $s \notin W$. It follows that W does not contain B^o. Thus, we have shown that $(E', \beta(E',E))$ is not bornological and hence not quasibarrelled.

35. A distinguished space whose strong dual is not separable.

The Banach space ℓ^1 is a distinguished space. But its strong dual ℓ^∞ is not separable.

36. A distinguished space whose strong dual is not metrizable.

The space w of all sequences equipped with the normal topology is a Fréchet space, but its strong dual $(w, \beta(w^\times , w)) = (\phi, \beta(\phi, w))$ is a non-metrizable barrelled space, where ϕ is the space of finite sequences. (To a sequence space λ, we define $\lambda^\times$ to be the sequence space consisting of all sequences $u = (u_i)$ for which the scalar product $ux = \sum_{i=1}^{\infty} u_i x_i$ converge absolutely for all $x \varepsilon \lambda$. For example $w^\times = \phi$, $\phi^\times = w$. Now let $\lambda \supset \phi$. Then λ and $\lambda^\times$

form a dual pair. Define the seminorms

$$p_u(x) = \sum_{i=1}^{\infty} |u_i||x_i| \ , \ u \in \lambda^{\times}$$

on λ. Then the family $\{p_u\}$ of semi-norms defines a locally convex topology on λ called normal topology. See B[20] , page 407).

37. A distinguished space which is not quasi-barrelled.

Let F be a non-reflexive Banach space (for example, $F = c_0$, the Banach space of sequences converging to 0). Let E be the Mackey space $(F', \tau(F', F))$. Then $E' = F$ and the strong topology on E' is the norm topology, because the $\sigma(F', F)$-bounded sets are norm-bounded. Thus, E is semi-reflexive but not reflexive so that E is not quasi-barrelled. (A locally convex space is reflexive iff it is semi-reflexive, and quasi-barrelled). Since E is semi-reflexive, it is distinguished.

38. A bornological space whose strong bidual is not bornological.

Let Λ be an infinite set and $\{f_i\}$ a sequence of positive functions on Λ such that $i<j$ implies $f_i \le f_j$. Then we obtain a bornological space E as the inductive limit of Banach spaces E_i whose unit sphere is $\{f \ ; \ |f(\lambda)| \le f_i(\lambda), \lambda \in \Lambda\}$. We show that E" is not bornological when functions f_i are adequately chosen. Let F be the totality of functions g on Λ such that

$$\sum_{\lambda \in \Lambda} f_i(\lambda) \ |g(\lambda)| < +\infty \ , \quad i = 1, 2, \ldots .$$

Then F can be considered as a subspace of E' (as functionals $f \to \sum_{\lambda \in \Lambda} f(\lambda) g(\lambda)$) and also as the dual of a subspace E_0 of E which consists of all functions $f \varepsilon E$ whose values are 0 except for a finite number of λ. If $\phi \varepsilon E'$ is continuous on E by the order topology, then $\phi \varepsilon F$. So, there is a continuous projection of E' onto F. Then there is also a continuous projection of E" onto F' and hence F' is a homeomorphic image of E". Therefore, it is sufficient, for our purpose, to choose f_i so as to make F' non-bornological. But this has already been found in ≠ 34 — Λ is the set of pairs (n, m) of positive integers and

$$f_i(n, m) = \begin{cases} m & \text{if } n \leq i \\ 1 & \text{if } n > i . \end{cases}$$

39. An (LB)-space which is not quasi-complete.

Let $E_0 = c_0$, with the elements written as double sequences $x = (x_{ij})$. $x \varepsilon E_0$ iff $\lim_{i,j \to \infty} |x_{ij}| = 0$. Clearly E_0 is a Banch space under the norm $||x|| = \sup_{i,j} |x_{ij}|$. Let

$$a_{ij}^{(n)} = \begin{cases} j & \text{if } i \leq n \\ 1 & \text{if } i > n . \end{cases}$$

Let E_n be the space of all double sequences with

$$\lim_{i,j \to \infty} \frac{|x_{ij}|}{a_{ij}^{(n)}} = 0$$

and norm

$$||x||_n = \sup_{i,j} \frac{|x_{ij}|}{a_{ij}^{(n)}} .$$

E_n is obtained from E_0 by making a diagonal transformation and so it is topologically isomorphic to c_0. The embedding $E_{n-1} \to E_n$ is continuous. The topological inductive limit $E = \lim E_n$, therefore exists, provided that the hull topology u on E is Hausdorff, and E is then an (LB)-space. If we put $a_{ij} = k$ for all i and j, the norm

$$||x||_\infty = \sup \frac{|x_{ij}|}{a_{ij}}$$

is weaker than $||x||_n$ on E_n , and so it defines a Hausdorff topology on E which is weaker than u. Thus, u is also Hausdorff. E' can be identified with the space of all double sequences $u = (u_{ij})$ for which

$$\sum_{i,j=1}^{\infty} |u_{ij}|\, a_{ij}^{(n)} < \infty, \quad n = 1, 2, \ldots \quad .$$

Let $e = (1, 1, \ldots)$. Let B be the set of all vectors $e^{(n)} = (e, \ldots , e, 0, 0, \ldots)$, e occuring in n places, $n = 1, 2, \ldots$. B is contained in E, and is weakly bounded. Since $e^{(n)} \varepsilon E_n - E_{n-1}$, B is not contained in any one of the spaces E_n . If E were quasi-complete, the closed circled convex cover of B, which would be complete, would have to lie in some E_n — because every circled convex bounded and complete subset of an (LB)-space $(E, u) = \bigcup_1^\infty (E_n, u_n)$ is a bounded subset of some (E_k, u_k) [20] — a fortiori, the same would be true of B.

40. A locally convex space which is not reflexive (not even semi-reflexive) but its strong dual is reflexive.

Let E be a dense proper subspace of an infinite-dimensional reflexive Banach space with the relative norm topology. Then E' is reflexive while E is not semi-reflexive.

41. A countably barrelled space which is not barrelled.

(i) It is a well-known fact (Cf. B [20] or B [15]) that the strong dual of a metrizable locally convex space is a complete (DF)-space ; a (DF)-space is countably quasi-barrelled and a complete countably quasibarrelled space is countably barrelled. But it need not be quasi-barrelled (and so neither barrelled nor bornoligical) as shown in $\#$ 34.

(ii) Let W be the space of ordinals less than the first uncountable ordinal. Then the space C(W) of all continuous (real or complex) functions on W, equipped with the compact-open topology is a countably barrelled space by Theorem 13, because the closure of a countable union of compact subsets of W is compact. But C(W) is not barrelled, since W is pseudo-compact but not compact.

42. A locally convex space C(X) of continuous functions which is not countably barrelled.

Let W^* be the space of ordinals less than or equal to the first uncountable ordinal, and let T be the Tychonoff plank. Then

$$T = \overline{\bigcup_{n=1}^{\infty} (W^* \times \{n\})}.$$

Clearly, each $W^* \times \{n\}$ is compact and T is pseudo-compact but not compact. Hence C(T) is not countably barrelled.

43. A semi-reflexive countably barrelled space which is

not barrelled.

Let Λ be an uncountable set. If E is the direct sum $\mathbb{R}^{(\Lambda)}$ and E' is the subspace of the product $\mathbb{R}^{\Lambda}$ consisting of all $x = (\zeta_\lambda)_{\lambda\varepsilon\Lambda}$ for which atmost countably many ζ_λ are non-zero, $< E,E' >$ is a dual pair. Given an atmost countable non-empty subset τ of Λ and a family $(\alpha_\lambda)_{\lambda\varepsilon\tau}$ of real numbers, the set

$$\{(\zeta_\lambda)_{\lambda\varepsilon\Lambda} \; ; \; |\zeta_\lambda| \leq \alpha_\lambda \text{ for } \lambda\varepsilon\tau \; , \quad \zeta_\lambda = 0 \text{ otherwise } \}$$

is $\sigma(E', E)$-compact, circled and convex. The topology on E of uniform convergence on all such sets is a topology of the dual pair $< E,E' >$, under which E is countably barrelled. Also $(E, \tau(E,E'))$ is semi-reflexive. $(E, \tau(E,E'))$ is not barrelled, since

$$E' \cap \prod_{\lambda\varepsilon\Lambda} [0,1]$$

is $\sigma(E',E)$-closed and bounded but is not $\sigma(E',E)$-compact.

44. A countably quasi-barrelled (and hence σ-quasi-barrelled) space which is not σ-barrelled.

The space ϕ of # 16 is a quasi-barrelled space and hence countably quasi-barrelled. But it is not σ-barrelled, because a quasi-barrelled σ-barrelled space is barrelled and ϕ is not barrelled as shown in # 16.

45. A σ-barrelled space which is not a Mackey space.

Let X be an uncountable vector space, and we denote

by $\mathcal{E}$ (respectively, $\mathcal{F}$) the vector space of real-or complex-valued functions on X which are different from 0 for a finite (respectively, an infinite countable) number of elements of X. Then the spaces $\mathcal{E}$ and $\mathcal{F}$ form a dual pair separating points which permits us to consider the weak space F associated to $\mathcal{F}$. We denote by E the space $\mathcal{E}$ equipped with the topology generated by the family of semi-norms

$$p_{\mathcal{B}}(f) = \sup\{|\phi(f)| \; ; \; \phi\in\mathcal{B}\}, \; f\in\mathcal{E},$$

where $\mathcal{B}$ varies over the countable family of bounded subsets of F. We note that for each $\mathcal{B}$, there is a countable subset $X_{\mathcal{B}}$ of X such that $p_{\mathcal{B}}(f) = 0$ for each $f\in E$ which is zero on $X_{\mathcal{B}}$. We can immediately deduce that each neighbourhood of 0 in E contains a nontrivial subspace of E. We show that E is σ-barrelled. For this, it is sufficient to show that $F = (E', \sigma(E', E))$. Now, if ϕ is a continuous linear functional on E, there exists a $\mathcal{B}$ such that

$$|\phi(\cdot)| \leq p_{\mathcal{B}}(\cdot).$$

We, then, have $\phi(f) = 0$ for each $f\in E$, which is zero on $X_{\mathcal{B}}$. The conclusion now follows. Next, we show that E is not a Mackey space. For this, it is sufficient to prove that

$$\theta = \{f\in E; \; |f(x)| \leq 1, \text{ for each } x\in X\}$$

is a neighbourhood of 0 equipped with the Mackey topology, for θ contains no nontrivial subspace of E. Now, if ϕ is a linear functional on E which is not continuous, there exists $\varepsilon>0$ such that $|\phi(x)|\geq\varepsilon$ for an infinite set of points $x\in X$. The conclusion now follows from the theorem of bipolars since θ is a barrel in E.

45(a). A σ-barrelled space which is not countably quasi-barrelled (and hence not countably barrelled).

Let the space X of # 45 be, in addition, a countable union of uncountable sets X_n which are increasing and such that $X \setminus X_n$ is uncountable for some $n \in \mathbb{N}$. We define E_1 as the space $\mathcal{E}$ endowed with the topology given by the system of semi-norms obtained by filtering those of E and the semi-norms

$$p_n(f) = \sup \{|f(x)| \ ; \ x \in X_n\}, \ f \in \mathcal{E} .$$

We show that E_1 is σ-barrelled: Note that F is also equal to $(E_1', \sigma(E_1', E))$. Infact, for each $n \in \mathbb{N}$, we have

$$\theta \subset b_n = \{f \in E_1 \ ; \ p_n(f) \leq 1\} ,$$

and so each linear functional bounded on b_n is bounded on θ, and hence continuous on E. The conclusion then follows from the following result: If G is σ-barrelled, it is σ-barrelled for each system P of semi-norms lying between those of G and those of Mackey space associated to G. (To prove this, we observe that each bounded sequence of $(G_p', \sigma(G_p', G_p))$ is a bounded sequence of $(G', \sigma(G', G))$ and hence is equicontinuous in G and, a fortiori, in G_p). Now we show that E_1 is not countably quasi-barrelled. Denote by χ_n $(n \in \mathbb{N})$ the set of characteristic functions of points of X_n. Each χ_n is clearly equicontinuous on E_1 and $\chi = \cup X_n$ is bounded in $(E_1', \beta(E_1', E))$ because χ is included in θ^0, that is, in a compact, circled and convex subset of $(E_1', \sigma(E_1', E))$. However, χ is not equicontinuous on E_1.

46. A Mackey space which is not σ-quasi-barrelled.

The Mackey space $E = (\ell^\infty, \tau(\ell^\infty, \ell^1))$ has property (S) but not property (C) as shown in # 25. Since a σ-barrelled space has property (C), E is not σ-barrelled. If E were σ-quasi-barrelled, then, since $(E', \sigma(E', E))$ is sequentially complete, E would be σ-barrelled which is not true.

47. A locally convex space which has property (C), but is not σ-barrelled.

Let E be an infinite-dimensional vector space with algebraic dual E^*. Then $(E, \sigma(E, E^*))$ has property (C), because every $\sigma(E^*, E)$-bounded subset of E^* is relatively $\sigma(E^*,E)$-compact. To show that $(E, \sigma(E,E^*))$ is not σ-barrelled, let B be any Hamel basis of E. Let B_0 be a countably infinite subset of B. For $x\varepsilon B_0$, let f_x be that linear functional in E^* such that

$$f_x(x) = 1$$

and

$$f_x(y) = 0, \ y\varepsilon B-\{x\} .$$

The set

$$A = \{f_x \ ; \ x\varepsilon B_0\}$$

is a countable $\sigma(E',E)$-bounded subset of E^*. A^0 contains no finite-codimensional subspace of E. Then A is not equicontinuous and so $(E,\sigma(E,E^*))$ is not σ-barrelled.

48. A sequentially barrelled space which is not σ-quasi-barrelled (and hence not σ-barrelled).

It is known that the space ℓ^∞ is a perfect sequence space and ℓ^1 its Köthe dual. Then $(\ell^\infty, \tau(\ell^\infty, \ell^1))$ is sequentially barrelled, because a Köthe function space Λ is sequentially barrelled in the Mackey topology $\tau(\Lambda,\Lambda^*)$-

In a Köthe function space Λ, each Cauchy sequence for the weak topology is weakly convergent P[15] — and hence the same is true of a perfect sequence space. But $(\ell^\infty, \tau(\ell^\infty, \ell^1))$ is not σ-quasibarrelled as shown in # 46.

49. A sequentially barrelled space which does not have property (S).

$(\ell^1, \tau(\ell^1, c_0))$ is a sequentially barrelled space, because a perfect sequence space with Mackey topology is sequentially barrelled (P[92]). Clearly, it does not have property (S).

50. A quasi-complete locally convex space which is not sequentially barrelled.

Let w be the space of all sequences which is a perfect sequence space. Let $\mathfrak{S}$ be the set of all sequences (a_i) such that $a_i = 0$ or 1 and

$$\frac{1}{n} \sum_{i=1}^{n} a_i \to 0 \quad \text{as } n \to \infty.$$

Define

$$w_0 = \{ax = (a_i x_i) \; ; \; a = (a_i) \varepsilon \mathfrak{S}, \; x = (x_i) \varepsilon w\} .$$

Denote by $|\sigma|(w^*, w)$ the normal topology on w^*. Then we have

(A) : $\beta(w^*, w_0) = \beta(w^*, w)$;

(B) : $\sigma(w^*, w_0) \leq |\sigma|(w^*, w_0) \leq \tau(w^*, w_0)$;

Hence a subset of w^* is $\sigma(w^*, w_0)$-bounded iff it is $|\sigma|(w^*, w_0)$-bounded.

(C) : Every $|\sigma|(w^*, w_0)$ - bounded subset of w^* is $|\sigma|(w^*, w)$-bounded. (For proofs, see P[92]).

Let ϕ be the space of all finite sequences. Then

$\sigma(\phi, w)$-bounded subsets of ϕ are finite-dimensional; hence by (B) and (C) the $\sigma(\phi, w_0)$-bounded subsets of ϕ are also finite-dimensional. Thus $(\phi, \tau(\phi, w_0))$ is quasi-complete. Now $(w_0, \sigma(w_0, \phi))$ is a barrelled metrizable space with completion w and bidual w by (A). Hence $(\phi, \tau(\phi, w_0))$ is not sequentially barrelled (because, if E is a metrizable locally convex space with $(E', \tau(E', E))$ sequentially barrelled, then E is complete [92]).

51. A (DF)-space which is not countably barrelled.

The space ϕ of # 16 is a normed space and hence a (DF) space. But it is not σ-barrelled as shown in # 44, and hence it is not countably barrelled either.

52. A (DF) space which is not quasi-barrelled.

Cf. # 41 (i).

53. A quasi-barrelled (DF)-space which is not bornological.

This counterexample is based on the following result:

(*) Let λ be an echelon space defined by the increasing system $\alpha^{(k)}$, $k = 1,2,\ldots$, such that every $\alpha^{(k)} = \{a_n^{(k)}\}_{n=1}^{\infty}$ is a sequence of strictly positive numbers.
Suppose that, for every positive integer p, there is a strictly increasing sequence $\{p_n\}_{n=1}^{\infty}$ of positive integers such that

(1) $\{a_{p_n}^{(k)} / a_{p_n}^{(1)}\}_{n=1}^{\infty}$ is a bounded sequence for $k = 1,2,\ldots, p$.

(2) $\lim_{n\to\infty} (a_{p_n}^{(p+1)} / a_{p_n}^{(1)}) = \infty$.

Then there is a dense subspace E in $(\lambda^{\times}, \sigma(\lambda^{\times}, \lambda))$ such that

$(E,\beta(\lambda^{\times},\lambda))$ is a non-bornological quasi-barrelled space.

This result as well as the counter-example that follows are due to M.Valdivia (Math. Z. 136 (1974), 249-251). For information about echelon space λ and co-echelon space $\lambda^{\times}$, See B [20] , page 419. For the example, we proceed as follows:

We choose a sequence $\alpha^{(k)} = \{a_n^{(k)}\}_{n=1}^{\infty}$ defined in the following way. Given a positive integer n, we write $n = (2m-1)2^{h-1}$, m and h being positive integers. If $h \leq k$, we put $a_n^{(k)} = (2m-1)2^{h-1}$. If $h > k$, then $a_n^{(k)} = 1$. Obviously, the system $\alpha^{(k)}$, k=1,2,..., is increasing. Given a positve integer p, we put $p_n = (2n-1)2^p$, n = 1,2,... . Then $a_{p_n}^{(k)} = 1$, k=1,2,...,p and $a_{p_n}^{(p+1)} = (2n-1)2^p$. Hence,

$$a_{p_n}^{(k)} / a_{p_n}^{(1)} = 1,\ k = 1,2,\ldots,p \quad \text{and} \quad \lim_{n\to\infty} (a_{p_n}^{(p+1)} / a_{p_n}^{(1)}) = \infty$$

and, therefore, conditions (1) and (2) of the result (*) are satisfied.

Now, we apply(*)to obtain a space $(E, \beta(\lambda^{\times}, \lambda))$ which is quasibarrelled and non-bornological. Since there is a countable fundamental system of bounded sets in $(E, \beta(\lambda^{\times},\lambda))$, $(E, \beta(\lambda^{\times}, \lambda))$ is a (DF)-space.

54. A locally topological space which is neither a bornological space nor a (DF)-space.

Cf. Adasch, N., Ernst, B. and Keim, D.: Topological vector spaces, Lecture Notes in Mathematics, 639 (1978), Springer-Verlag, Berlin-Heidelberg-New York.

(A locally topological space which is a common generalization of bornological and (DF)-spaces is defined as follows: Let β be the family of all bounded and circled sets in a locally convex space (E, u) such that

$B_1, B_2 \varepsilon \mathcal{B} \Rightarrow B_1 \cap B_2, B_1 + B_2 \varepsilon \mathcal{B}$. A linear map f: E→F, F a locally convex space, is called locally continuous if all restrictions f/B, $B \varepsilon \mathcal{B}$, are continuous at 0 in the topology induced by u on B. (E,u) is called a locally topological space if all locally continuous maps from E into any locally convex space F are continuous).

55. A k-quasibarrelled space which is not k-barrelled.

Let E be the vector space of complex sequences with only a finite number of non-zero elements, equipped with the topology of pointwise covergence. Then E is metrizable and so quasibarrelled which implies that E is k-quasi-barrelled. We now show that E is not k-barrelled for any $k \geq 1$. Consider

$$U_n = \{x \varepsilon E \ ; \ |x_j| \leq 1, \ \ 1 \leq j \leq n\} \ .$$

Each U_n is a circled, convex and closed neighbourhood of 0 in E. Their intersection is a barrel in E. Let

$$U(k) = \bigcap_{n=1}^{\infty} \{k^n U_n\} \ , \ k \geq 1 \ .$$

Let

$$a_n = e^{e^n}, \ a^{(n)} = (a_1, a_2, \ldots, a_n, 0, 0, \ \ldots \ ).$$

Then $B = \{a^{(n)} \ ; \ n = 1,2, \ldots\}$ is a bounded subset of E that is not absorbed by U(k) for any $k \geq 1$. Hence E is not k-barrelled for any $k \geq 1$.

56. An H-space which is not a distinguished space.

Cf. # 34.

57. An H-space which is not metrizable.

Let ϕ be the space of all finite sequences and u the normal topology on ϕ. Then it is the same as the locally convex direct sum topology on ϕ and hence (ϕ, u) is not metrizable (See # 11). But (ϕ, u) is a distinguished space, because $(\phi^{\times}, \beta(\phi^{\times},\phi)) = (w, \beta(w,\phi))$ is a Fréchet space. Hence (ϕ, u) is an H-space.

58. An H-space whose strong dual is not separable.

Cf. # 34.

....

OPEN PROBLEMS

1. Do the concepts of Baire and unordered Baire-like locally convex spaces coincide?

2. Is every countable-codimensional subspace of a Baire space of the same kind? (This is satisfied by unordered Baire-like spaces — Math. Ann. 206(1973), 23-34. So, the invalidity of this problem would distinguish between Baire and unordered Baire-like spaces).

3. (I.Tweddle): Is there a locally convex space E such that $(E, \tau(E, E'))$ is countably barrelled but not barrelled?

4. Is there a bornological space whose strong dual is not countably barrelled?

5. (Levin and Saxon P[56]): Is a Mackey space with property (C) always a σ-barrelled space?

6. If F is a σ-barrelled space, is it necessarily

countably barrelled under the topology of uniform convergence on the $\sigma(F',F)$-bounded separable sets?

7. (Crothendieck): Is there a barrelled (DF) space which is not bornological?

CHAPTER 4

SPECIAL CLASSES OF TOPOLOGICAL VECTOR SPACES

Introduction

A topological space X which satisfies the condition of Theorem 4 in the introduction to Chapter 3 is referred to as an N-S space. A topological space X is said to be an L-W space if for every decreasing sequence $\{F_n\}$ of closed and noncompact subsets of X for which $\cap F_n \neq \emptyset$, there exists $f \in C(X)$ unbounded on each F_n. C(X) is a barrelled space whenever X is an L-W space.

A topological vector space E is W-barrelled if every closed, circled and absorbing set in E is a neighbourhood of 0.

THEOREM 1. Let X be a topological space which is first countable at a point p which has no compact neighbourhood. Then X is not an L-W space. (Cf. P[55])

THEOREM 2. If C(X) is W-barrelled, then X is an L-W space. (Cf. P[55])

A closed and circled subset B of a topological vector space is called an ultrabarrel (a bornivorous ultrabarrel) if there exists a sequence $\{B_n\}$ of closed, circled and absorbing (closed, circled and bornivorous) subsets of E such that $B_1 + B_1 \subseteq B$ and $B_{n+1} + B_{n+1} \subseteq B_n$ for all $n \geq 1$. The sequence $\{B_n\}$ is called a defining sequence for B. B is called a suprabarrel (bornivorous suprabarrel) if the closedness of B and B_n, for each n, is dropped from the above definition.

A topological vector space E is called ultrabarrelled (quasi-ultrabarrelled, ultrabornological) if each ultrabarrel (respectively, bornivorous ultrabarrel, bornivorous suprabarrel) in E is a neighbourhood of 0. Ultrabarrelled and ultrabornological spaces are quasi-ultrabarrelled. A locally convex ultrabarrelled (quasi-ultrabarrelled, ultrabornological) space is barrelled (quasibarrelled, bornological).

If (E,u) is a topological vector space, we write u^{00} to denote the finest locally convex topology coarser than u. If (E,u) is metrizable, so is (E,u^{00}) and so, if $< E, E' >$ is a dual pair, $u^{00} = \tau(E,E')$.

PROPOSITION 1. If (E,u) is a metrizable topological vector space with dual E', $< \tilde{E}_u, E' >$ is a dual pair and (E,u^{00}) is ultrabarrelled, then $u = u^{00}$. (Cf. P[74])

For each integer $k \geq 1$, let $\{V_k^{(n)} ; n = 0,1,2 \ldots\}$ be a sequence of closed and circled neighbourhoods of 0 in a topological vector space E such that

$$V_k^{(n+1)} + V_k^{(n+1)} \subseteq V_k^{(n)}$$

for all n. If, for each n,

$$V^{(n)} = \bigcap_{k=1}^{\infty} V_k^{(n)}$$

is absorbing (bornivorous), then $V^{(0)}$ is an ultrbarrel (a bornivorous ultrabarrel) in E with $\{V^{(n)} ; n = 1,2\ldots\}$ as a defining sequence. $V^{(0)}$ is called an ultrabarrel (a bornivorous ultrabarrel) of type (α). A topological vector space E is called a countably ultrabarrelled (countably quasi-ultrabarrelled) space if each ultrabarrel (bornivorous ultrabarrel) of type (α) is a neighbourhood of 0 in E. Every ultrabarrelled (quasi-ultrabarrelled) space is countably ultrabarrelled (countably quasi-ultrabarrelled) and every countably ultrabarrelled space is countably quasi-ultrabarrelled. A locally convex space which is countably ultrabarrelled (countably quasi-ultrabarrelled) is countably barrelled (countably quasibarrelled).

A topological vector space E is called k-ultrabarrelled (k-quasi-ultrabarrelled) if $\bigcap_{j=1}^{\infty} k^j V_j^{(0)}$ is a neighbourhood of 0 whenever $V^{(0)} = \bigcap_{j=1}^{\infty} V_j^{(0)}$ is an ultrabarrel (bornivorous ultrabarrel) of type (α). A k-ultrabarrelled space is k-quasi-ultrabarrelled. A locally convex space which is k-ultrabarrelled (k-quasi-ultrabarrelled) is k-barrelled (k-quasi-barrelled).

A locally semiconvex space E is called hyperbarrelled (quasi-hyperbarrelled) if every closed, circled, semiconvex and absorbing (bornivorous) set in E is a neighbourhood of 0. E is hyperbornological if every circled, semiconvex and absorbing set in E is a neighbourhood of 0. Clearly hyperbarrelled and hyperbornological spaces are quasi-hyperbarrelled. E is called γ-hyperbarrelled (γ-quasi-hyperbarrelled) if every closed, circled, semiconvex and absorbing (bornivorous) set V in E is a neighbourhood of 0 whenever it satisfies the following condition : $V = \bigcap_{\alpha \in \Phi} V_\alpha$, where, for some $\lambda>0$, each V_α is closed, circled and λ-convex neighbourhood of 0 and the cardinal of Φ is γ. (V_α is λ-convex if $V_\alpha + V_\alpha \subseteq \lambda V_\alpha$).

A hyperbarrelled (quasi-hyperbarrelled, $\mathcal{N}$-hyperbarrelled) space is $\mathcal{N}$-hyperbarrelled ($\mathcal{N}$-quasi-hyperbarrelled, $\mathcal{N}$-quasi-hyperbarrelled) for every $\mathcal{N}$.

THEOREM 3. If $\{f_n\}$ is a pointwise convergent sequence of continuous linear maps from an $\mathcal{N}_0$-hyperbarrelled space E into a locally semiconvex space F, then its limit map is continuous. If, in addition, F is sequentially complete, $\{f_n\}$ is necessarily pointwise convergent if it is pointwise bounded on a set which is everywhere dense in E. (Cf. P[35])

Let f be a linear map from a topological vector space E into another topological vector space F. We say that the filter condition holds if for any Cauchy filter basis $\mathfrak{F}$ on E such that f($\mathfrak{F}$) is convergent to a point of f(E), it follows that $\mathfrak{F}$ is convergent to an element in E. If E is complete, the filter condition holds. Let E be a topological vector space under each of two topologies u and v. We say that the closed neighbourhood condition holds if there is a basis υ of u-neighbourhoods of 0 which are v-closed.

THEOREM 4. Let (F,v) be a locally convex space. Then F is barrelled iff the only convex topologies with bases of v-closed neighbourhoods of 0 are those coarser than v. (Cf. B[15])

1. A topological vector space in which the filter condition holds but not the closed neighbourhood condition.

Let (E,u) be a Banach space with $E' \neq E^*$, (Any infinite dimensional Banach space satisfies this condition).

Let $v = \tau(E,E^*)$. Then (E,v) is complete and so the filter condition holds; but the closed neighbourhood condition does not hold by Theorem 4, because v is strictly finer than u.

2. An N-S space which is not an L-W space.

The metric space Q of all rational numbers is clearly an N-S space. But it is not an L-W space in view of Theorem 1.

3. A locally convex space $C(X)$ of continuous functions which is barrelled and bornological but not W-barrelled.

Let $X = Q$ be the metric space of all rational numbers. Since it is a Lindelöf space, it is realcompact (Q-space) so that $C(X)$ is bornological. Further, since X is an N-S space, $C(X)$ is a barrelled space. But $C(X)$ is not W-barrelled in view of Theorem 2.

4. An ultrabarrel which is not convex and which does not have a defining sequence of convex sets.

Let E be a complete locally bounded space which is not locally convex (say, $E = L^p[0,1]$, $0<p<1$). Let B be a closed, circled and bounded neighbourhood of 0. Then B is an ultrabarrel with $\{\lambda_n B\}$ as a defining sequence for some sequence $\{\lambda_n\}$ of positive real numbers. But B and $\lambda_n B$ are not convex.

5. An ultrabarrelled space which is not barrelled.

The topological vector space $E = L^p[0,1]$, $0<p<1$, (See Chapter 1, # 3(i)) is complete and metrizable and hence an ultrabarrelled space. Clearly it is not barrelled.

6. A barrelled space which is not ultrabarrelled.

Consider the sequence space

$$\ell^{\frac{1}{2}} = \{x = (x_n) ; \sum_{n=1}^{\infty} |x_n|^{\frac{1}{2}} < \infty\}.$$

Let

$$||x||_{\frac{1}{2}} = (\Sigma |x_n|^{\frac{1}{2}})^2.$$

Let u be the topology with a basis of neighbourhoods of 0 formed by the sets $\{x ; ||x||_{\frac{1}{2}} \leq \varepsilon\}$. Then $(\ell^{\frac{1}{2}}, u)$ is a complete and metrizable topological vector space and so ultrabarrelled. Hence $(\ell^{\frac{1}{2}}, u^{00})$ is barrelled. Now, let

$$||x||_1 = \Sigma |x_n|.$$

Then

$$||x||_1 \leq ||x||_{\frac{1}{2}},$$

$\ell^{\frac{1}{2}}$ is a subspace of ℓ^1 and the topology induced on $\ell^{\frac{1}{2}}$ by the norm topology of ℓ^1 is coarser than u. The space ϕ of all finite sequences is dense in ℓ^1 under the norm topology, and so, since $\phi \subseteq \ell^{\frac{1}{2}}$, $\ell^{\frac{1}{2}}$ is dense in ℓ^1 under the norm topology. Hence the dual of $\ell^{\frac{1}{2}}$ under the norm topology is ℓ^{∞} and the norm topology is thus $\tau(\ell^{\frac{1}{2}}, \ell^{\infty})$. The dual of $(\ell^{\frac{1}{2}}, u)$ is also ℓ^{∞} and so $u^{00} = \tau(\ell^{\frac{1}{2}}, \ell^{\infty})$. By proposition 1, it follows that $(\ell^{\frac{1}{2}}, u^{00})$ is not ultra-barrelled.

7. An u^{00}- compact set which is not u-compact.

We construct a subset of $\ell^{\frac{1}{2}}$ which is u-precompact but whose circled and convex envelope is not u-precompact, and deduce that a u^{00}-compact set is not u-compact. Let

$$x_{11} = (1,0,0, \dots)$$
$$x_{21} = (0,\tfrac{1}{2},0, \dots)$$
$$x_{22} = (0,0,\tfrac{1}{2},0, \dots)$$
$$x_{31} = (0,0,0,1/3,0, \dots)$$
$$x_{32} = (0,0,0,0,1/3,0, \dots)$$
$$x_{33} = (0,0,0,0,0,1/3,0, \dots)$$

and in general let x_{nm} $(1 \leq m \leq n)$ be the element with all terms zero except the $(\frac{1}{2}\, n(n-1)+m)^{th}$ term which is $1/n$. Let

$$B = \{x_{nm} \ ; \ n,m = 1,2, \dots\} .$$

Then

$$||x_{nm}||_{\frac{1}{2}} = \frac{1}{n}$$

and $x_{nm} \to 0$, regarding (x_{nm}) as the sequence x_{11}, x_{21}, x_{22}, x_{31}, ... Hence B is u-precompact. Define the sequence (y_n) by

$$y_n = \frac{1}{n} \sum_{m=1}^{n} x_{nm}$$

so that y_n is the element with the first $\frac{1}{2}$ $n(n-1)$ terms zero, then n terms each equal to $\frac{1}{n^2}$ and the remaining terms zero. Then y_n belongs to the circled and convex envelope of B; if $n \neq k$,

$$||y_n - y_k||_{\frac{1}{2}} = 4$$

and so (y_n) is not u-compact. Hence the circled and convex envelope of B is not u-precompact. Also $||y_n|| = \frac{1}{n}$ and

$$A = \{0\} \cup \{y_n \ ; \ n = 1,2, \dots\}$$

is u^{00}-compact; but certainly A is not u-compact.

8. An ultrabarrelled space which is not non-meagre.

The space ϕ of finite (real) sequences, under the finest compatible topology u is ultrabarrelled. Since ϕ is a countable union of finite-dimensional spaces, it is meagre.

9. An ultrabornological space which is not bornological.

The space $\underline{L}^p[0,1]$, $0<p<1$, as defined in Chapter 1, # 3(i), is a complete and metrizable topological vector space and hence it is ultrabornological. Clearly it is not bornological.

10. A bornological space which is not ultrabornological.

Let E be a vector space of uncountable dimension. Then the finest linear topology s is strictly finer than $\tau(E,E^*)$ (P[48], Theorem 3.1), so that the identity map of $(E, \tau(E,E^*))$ into (E,s) is not continuous; but it is bounded, since every $\tau(E,E^*)$-bounded subset is contained in a finite-dimensional subspace of E. Hence $(E, \tau(E,E^*))$ is not ultrabornological thought it is bornological.

11. An ultrabornological space which is not ultrabarrelled.

Let E be a countably dimensional non-locally convex metrizable topological vector space. Since it is metrizable, it is ultra-bornological. Since the finest linear topology on a countably dimensional vector space is locally convex, any ultrabarrelled topology on a countably dimensional vector space is necessarily locally convex. Hence E is not ultrabarrelled.

12. An ultrabarrelled space which is not ultrabornological.

Let E be an incomplete Hausdorff inductive limit of a sequence of Banach spaces, $\tilde{E}$ the completion of E and $x \varepsilon \tilde{E}-E$. Then the subspace E_1 of $\tilde{E}$ spanned by E and x is ultrabarrelled (P[36]). But E_1 is not bornological (P[49], page 155) and hence not ultrabornological.

13. A quasi-ultrabarrelled space which is neither ultrabarrelled nor ultra-bornological.

Let E_1 be an ultrabarrelled space which is not ultrabornological and E_2 an ultrabornological space which is not ultrabarrelled. Then the product of E_1 and E_2 is quasi-ultrabarrelled which is neither ultrabarrelled nor ultrabornological.

14. A countably quasi-ultrabarrelled space which is not countably ultrabarrelled.

The normed space E of Chapter 3, # 16 is not countably barrelled (Chapter 3, # 44) and hence not countably ultrabarrelled either. However, it is quasi-ultrabarrelled, being a normed space, and hence countably quasi-ultrabarrelled.

15. A countably ultrabarrelled space which is not ultrabarrelled.

The strong dual of a metrizable locally convex space is countably ultrabarrelled (P[37], Proposition 3.1) and so the space considered on page 435 in B[20] is countably ultrabarrelled. But, as shown there, it is not barrelled and hence not ultrabarrelled either.

16. A countably barrelled space which is not countably

ultrabarrelled.

Let (E,u) be the complete metrizable topological vector space as defined in # 6. If $\tau(E,E^*)$ is the finest locally convex topology on E, then $(E, \tau(E,E^*))$ is barrelled and hence countably barrelled. We show that it is not countably quasi-ultrabarrelled: For each integer n and element $x = (x_1, x_2, \ldots)$ in E, let

$$t_n(x) = (x_1, x_2, \ldots , x_n, 0,0,\ldots).$$

Then (t_n) is a sequence of continuous linear maps from (E,u^{00}) into (E,u) such that, for each x in E, $t_n(x)$ converges to x in (E,u). Clearly, each t_n is a continuous map from $(E, \tau(E,E^*))$ into (E,u). Moreover, $\{t_n\}$ is uniformly bounded on the $\tau(E,E^*)$-bounded subsets of E. For, any $\tau(E,E^*)$-bounded subset A of E is contained in some finite-dimensional subspace E_0 of E, and the restrictions of $\{t_n\}$ to E_0 must be equicontinuous, implying that $\cup t_n(A)$ is u-bounded. However, since the identity map from $(E, \tau(E,E^*))$ onto (E,u) is not continuous, $\{t_n\}$ is not an equicontinuous set of maps from $(E, \tau(E,E^*))$ into (E,u). Hence, by (P[37], Theorem 3.2), $(E, \tau(E,E^*))$ is not countably quasi-ultrabarrelled and, a fortiori, not countably ultra-barrelled.

17. A k-quasi-ultrabarrelled space which is not k-ultra-barrelled.

The space E of complex sequences with only a finite number of non-zero elements equipped with the topology of pointwise convergence is k-quasi-ultrabarrelled, being metrizable. But it is not a k-ultrabarrelled space, because it is locally convex but not k-barrelled (Chapter 3, # 55).

18. A hyperbarrelled space which is not hyperbornological.

The space E_1 as defined in # 12 is locally convex ultrabarrelled and hence hyperbarrelled. Since a convex hyperbornological space is bornological and since E_1 is not bornological, it follows that E_1 is not hyperbornological.

19. A hyperbornological space which is not hyperbarrelled.

Let E be a countably infinite dimensional normed space. Then E is not barrelled and so not hyperbarrelled either. However, since E is metrizable, it is hyperbornological.

20. A quasi-hyperbarrelled space which is neither hyperbarrelled nor hyperbornological.

Let E be a hyperbarrelled space which is not hyperbornological and F a hyperbornological space which is not hyperbarrelled. Then E×F is a quasi-hyperbarrelled space which is neither hyperbarrelled nor hyperbornological.

21. An $\aleph$-quasi-hyperbarrelled space which is not $\aleph$-hyperbarrelled.

The space $E = \phi$ of Chapter 3, # 16, is, for each $\aleph \geq \aleph_0$, an $\aleph$-quasi-hyperbarrelled space which is not $\aleph$-hyperbarrelled.

22. A barrelled space which is not $\aleph_0$-hyperbarrelled.

Let (E,u) and (E,u^{00}) be as in # 6. Then $\{t_n\}$, as defined in # 16, is a sequence of continuous linear maps from (E,u^{00}) into (E,u) such that, for each x in $E, t_n(x)$ converges to x in (E,u). Since the identity map from (E,u^{00}) into (E,u) is not continuous, it follows from Theorem 3 that the barrelled space (E,u^{00}) is not γ_0-hyperbarrelled.

CHAPTER 5

ORDERED TOPOLOGICAL VECTOR SPACES

Introduction

All vector spaces in this Chapter are over the field $\mathbb{R}$ of real numbers.

An ordered vector space (E,C) which is also a topological vector space is called an ordered topological vector space; we denote it by (E,C,u) where u is the topology. The positive cone C in an ordered topological vector space (E,C,u) is called normal for the topology u if there is a neighbourhood basis $\mathcal{B}$ at 0 in (E,C,u) consisting of full sets. The members of $\mathcal{B}$ can be chosen to be circled. If u is a locally convex topology, we can assume that the members of $\mathcal{B}$ are convex.

THEOREM 1. Let (E,C,u) be an ordered topological vector space. The following statements are equivalent;

(a) C is normal.

(b) There is a neighbourhood basis $\mathcal{B} = \{V\}$ at 0 such that $0 \leq x \leq y$, $y \in V$ implies $x \in V$.

(c) For any two nets $\{x_\alpha\}$ and $\{y_\alpha\}$ in (E,C,u), if $0 \leq x_\alpha \leq y_\alpha$ for all α and if $\{y_\alpha\}$ converges to 0 for u, then $\{x_\alpha\}$ converges to 0 for u. (Cf. B[26])

PROPOSITION 1. Let (E,C,u) be an ordered topological space with normal cone C. Then each order-bounded subset of E is u-bounded. (Cf. B[26])

THEOREM 2. Let (E,C,u) be an ordered locally convex

space. Then the following statements are equivalent:

(a) C is normal.

(b_1) There is a family $\{p_\alpha\}$ of seminorms giving the topology u such that $0 \leq x \leq y$ implies $p_\alpha(x) \leq p_\alpha(y)$ for all α(equivalently, $p_\alpha(t+s) \geq p_\alpha(t)$ for all t, $s \varepsilon C$ and all α). (Cf. B[26])

COROLLARY 1. If C is a normal cone in an ordered locally convex space (E,C,u), so is $\overline{C}$.

THEOREM 3. Let (E, $||\cdot||$) be a space ordered by the positive cone C. The following statements are equivalent:

(a) C is normal for the norm topology.

(b) There is an equivalent norm $||\cdot||_1$ on E such that $||x||_1 \leq ||y||_1$ whenever $0 \leq x \leq y$.

(c) There is a constant $\lambda > 0$ such that $\lambda||x|| \leq ||y||$ whenever $0 \leq x \leq y$.

(d) There is a constant $\lambda > 0$ such that $||x + y|| \geq \lambda \max \{||x||, ||y||\}$ for all $x,y \varepsilon C$.

(e) The set $\{||x||;\ 0 \leq x \leq y,\ ||y|| \leq 1\}$ is bounded above. (Cf. B[26])

Let $\mathfrak{G}$ be the saturated class of all u-bounded subsets of an ordered locally convex space (E,C,u) such that $E = \bigcup_{S \varepsilon \mathfrak{G}} S$. The cone C in E is called a b-cone (strict b-cone) if the sets $\{\overline{S \cap C - S \cap C}\ ;\ S \varepsilon \mathfrak{G}\}$(respectively, $\{S \cap C - S \cap C\ ;\ S \varepsilon \mathfrak{G}\}$) form a fundamental system for $\mathfrak{G}$. We say that the cone C in an ordered topological vector space (E,C,u) gives an open decomposition in E if each

$V \cap C - V \cap C$ is an u-neighbourhood of 0 whenever V is an u-neighbourhood of 0.

THEOREM 4. Let u be a metric vector topology in an ordered vector space E such that C is generating and complete. If C gives an open decomposition in (E,u), then (E,u) is complete. (Cf. B[39])

An ordered topological vector space (E,C,u) is boundedly order-complete if every u-bounded and directed ($\leq$) subset has a supremum.

An ordered topological vector (ordered locally convex) space (E,C,u) which is a vector lattice is called a topological vector lattice (locally convex lattice) if there is a neighbourhood basis of 0 for u consisting of solid sets. Since the convex hull of a solid set in a vector lattice is solid, there is a neighbourhood basis of 0 for u consisting of convex and solid sets in a locally convex lattice and the topology is referred to as locally solid topology.

A normed space (Banach space) $(E, ||\cdot||)$ which is a vector lattice is called a normed vector lattice (Banach lattice) if $|x| \leq |y|$ implies $||x|| \leq ||y||$.

PROPOSITION 2. Let (E,C,u) be an ordered topological vector lattice which is a vector lattice. The following statements are equivalent:

(i) $(x,y) \to x \wedge y$ is a continuous map of $E \times E$ into E.

(ii) $(x,y) \to x \vee y$ is a continuous map of $E \times E$ into E.

(iii) $x \to x^+$ is a continuous map on E.

(iv) $x \to x^-$ is a continuous map on E.

(v) $x \to |x|$ is a continuous map on E. (Cf. B[26])

If the cone in E is normal, the continuity of each map in (iii), (iv) and (v) is equivalent to its continuity at 0. The maps defined in (i) through (v) are referred to as lattice operations.

THEOREM 5. A topological vector space (E,u) which is a vector lattice is a topological vector lattice iff the cone in E is normal and the lattice operations are continuous. (Cf. B[26])

THEOREM 6. Let (E,u) be a locally convex space which is a vector lattice. The following statements are equivalent.

(i) (E,u) is a locally convex lattice.

(ii) For any nets $\{x_\alpha\}$ and $\{y_\alpha\}$ in E, if $|x_\alpha| \leq |y_\alpha|$ for all α and $x_\alpha \to 0$ for u, then $y_\alpha \to 0$ for u.

(iii) There is a family $\{p_\alpha;\ \alpha \in I\}$ of seminorms on E generating u such that $|x| \leq |y|$ implies $p_\alpha(x) \leq p_\alpha(y)$ for all α. (Cf. B[26])

A vector sublattice M in a topological vector lattice (E,C,u) is a topological vector lattice in the relative topology. If M is a lattice ideal in E, then the quotient E/M is a topological vector lattice for the quotient topology. A product of topological vector lattices is a topological vector lattice.

An ordered normed space $(E,C,||\cdot||)$ is called an

order-unit normed space if there is an order-unit e such the norm $||\cdot||$ is the gauge of $[-e,e]$. $(E,C, ||\cdot||)$ is called an approximate order-unit normed space if there is an approximate order-unit $\{e_\lambda , \lambda\varepsilon\Lambda ,\leq\}$ in C such that the norm $||\cdot||$ is the gauge of the circled and convex set

$$S_\Lambda = \cup\{[-e_\lambda , e_\lambda]; \lambda\varepsilon\Lambda\}.$$

An order-unit normed space is clearly an approximate order-unit normed space.

THEOREM 7. Let $(E,C,||\cdot||)$ be an ordered normed space. The following statements are equivalent.

(i) E is an approximate order-unit normed space.

(ii) The open unit ball $B = \{x\varepsilon E ; ||x||<1\}$ is solid and directed upwards. (Cf. B[39])

Let (E,C) be an ordered vector space, where C is generating and let B be a base of C. The circled and convex envelope $\Gamma(B)$ of B is absorbing in E. The gauge of $\Gamma(B)$ is called a base semi-norm on E. If the seminorm is infact a norm then it is referred to as a base norm. An ordered normed space $(E,C, ||\cdot||)$ is called a base normed space if C is generating and if there is a base B of C such that the norm $||\cdot||$ is the gauge of $\Gamma(B)$.

PROPOSITION 3. If an ordered normed space $(E,C,||\cdot||)$ is an approximate order-unit normed space, then $(E',C', ||\cdot||)$ is a base normed space. (Cf. B[39])

The order-bound topology u_0 on an ordered vector space is the finest locally convex topology u for which every order-bounded set is u-bounded.

PROPOSITION 4. A neighbourhood basis at 0 for the order-bound topology u_0 on an ordered vector space is given by the family of all circled convex and order-bornivorous (a set which absorbs all order-bounded sets) sets. (Cf. B[26])

A locally convex space E which is a vector lattice is called a pseudo-M-space if it has a neighbourhood basis at 0 consisting of sublattices. A locally convex lattice which is a pseudo-M-space is called an M-space.

A locally convex lattice (E,C,u) is called order-quasibarrelled if each order-bornivorous barrel in E is a neighbourhood of 0.

Let (E,C,u) be a locally convex lattice. For any $x \in C$, we define

$$p_x(f) = \sup \{|\langle y,f \rangle| ; y \in [-x,x]\}$$

for all $f \in E'$. Then the family $\{p_x ; x \in C\}$ defines a locally solid topology on E' denoted by $\sigma_s(E',E)$.

THEOREM 8. Let (E,C,u) be a locally convex lattice. The following statements are equivalent.

(a) (E,C,u) is order-quasibarrelled.

(b) Each $\sigma_s(E',E)$-bounded subset of E' is equicontinuous. (Cf. B[15] or B[39])

A locally convex lattice (E,C,u) is called a countably order-quasi-barrelled vector lattice (abbreviated to C.O.Q. vector lattice) if each order-bornivorous barrel which is the countable intersection of closed, circled and convex neighbourhoods of 0 is itself a neighbourhood of 0 in E.

Every order-quasi-barrelled vector lattice is a C.O.Q. vector lattice. A countably barrelled locally convex lattice is a C.O.Q. vector lattice and a C.O.Q. vector lattice is a countably quasibarrelled locally convex lattice.

THEOREM 9. Let (E,C,u) be a locally convex lattice. The following statements are equivalent.

(a) (E,C,u) is a C.O.Q. vector lattice.

(b) Each solid barrel which is the countable intersection of closed, convex and solid neighbourhoods of 0 is itself a neighbourhood of 0 in E.

(c) Each $\sigma_S(E',E)$-bounded subset of E' which is the countable union of equicontinuous subsets of E' is itself equicontinuous. (Cf. B[15])

A.C.O.Q. vector lattice which has a fundamental sequence of order-bounded subsets is called an order-(DF) vector lattice.

A C.O.Q. vector lattice with an order-unit is an order-(DF) vector lattice. In particular an order-unit-normed vector lattice is an order-(DF) vector lattice.

Let (E,C,u) be a topological vector lattice. If, in the definition of an ultrabarrel (Chapter 4, Introduction), each B_n is circled and order-bornivorous (solid and absorbing), then B is said to be order-bornivorous ultra-barrel (solid ultrabarrel, if B is solid). (E,C,u) is called an order-quasi-ultrabarrelled vector lattice (abbreviated to O.Q.U. vector lattice) if each order-bornivorous ultrabarrel in E is a neighbourhood of 0.

THEOREM 10. A topological vector lattice (E,C,u) is an O.Q.U. vector lattice iff each solid ultrabarrel in E is a neighbourhood of 0. (Cf. B[15])

An ultrabarrelled topological vector lattice is an O.Q.U. vector lattice and an O.Q.U. vector lattice is a quasi-ultrabarrelled topological vector lattice. A locally convex O.Q.U. vector lattice is order-quasibarrelled. An order-unit-normed vector lattice is an O.Q.U. vector lattice.

Let (E,C,u) be a topological vector lattice. If, in the definition of an ultrabarrel of type (α) (Chapter 4, Introduction), $V^{(n)}$ is order-bornivorous then V^0 is called an order-bornivorous ultrabarrel of type (α) in (E,C,u). On the other hand, if each $V_k^{(n)}$ is closed and solid, then V^0 is called a solid ultrabarrel of type (α) in (E,C,u). (E,C,u) is called a countably O.Q.U. vector lattice if each order-bornivorous ultrabarrel of type (α) in (E,C,u) is a neighbourhood of 0.

THEOREM 11. A topological vector lattice (E,C,u) is a countably O.Q.U. vector lattice iff each solid ultrabarrel of type (α) is a neighbourhood of 0 in (E,C,u). (Cf. P[45])

THEOREM 12. Let (E,C,u) be a countably O.Q.U. vector lattice and (F,K,v) any topological vector lattice. If $\{f_n\}$ is a pointwise bounded sequence of continuous lattice homomorphisms of E into F, then $\{f_n\}$ is equicontinuous. (Cf. P[45])

An O.Q.U. vector lattice is a countably O.Q.U. vector lattice. A countably ultrabarrelled topological vector lattice is a countably O.Q.U. vector lattice which is a countably quasi-ultrabarrelled topological vector lattice. A locally convex lattice which is also a countably O.Q.U. vector lattice is a C.O.Q. vector lattice.

1. An ordered topological vector space with generating cone which does not give open decomposition.

Consider the Banach space $E = C[0,1]$ of all continuous (real-valued) functions on the closed interval $[0,1]$, equipped with the usual supremum norm. Let C be the cone in E consisting of all non-negative and convex functions and let $F = C - C$. By the Stone-Weierstrass theorem, F is dense in the Banach space E. Also, since any function in C must be differentiable on $[0,1]$ except at, atmost, a countable number of points, any function f in E which is not differentiable at an uncountable subset of $[0,1]$ is not in F. Therefore, F is a proper dense subspace of E, and F is a non-complete normed space in the relative topology. Since C is a generating and norm-complete cone in F, it follows from Theorem 4 that C does not give an open decomposition.

2. An ordered topological vector space with normal cone but with a (topologically) bounded set which is not order-bounded.

Let (S,Σ,μ) be an arbitrary measure space. Consider the space $E = L^p(S,\Sigma,\mu)$, $p \geq 1$, equipped with the norm

$$||f|| = \{\int |f(t)|^p \, d\mu(t)\}^{1/p} .$$

The cone of functions in E that are non-negative on S μ-almost everywhere is normal for the norm topology. However, the unit ball

$$B = \{f \varepsilon E \; ; \; ||f|| \leq 1\}$$

is norm-bounded but not order-bounded.

3. A cone in a topological vector space, which is not normal.

Let $E = \ell^\infty$ (or c_0 or ϕ), equipped with the supremum norm and P_s the cone consisting of sequences in E having all partial sums non-negative. We show that P_s is not normal in E. Let

$$x_n = e_1 + \ldots + e_n - ne_{n+1}$$

and

$$y_n = ne_{n+1}$$

where e_n is the element in E having 1 in the nth place and zeros elsewhere. Then

$$x_n \, , \, y_n \, \varepsilon P_s \, , \, ||x_n|| = ||y_n|| = n$$

and

$$||x_n + y_n|| = 1.$$

Hence P_s is not normal in E, in view of Theorem 3.

4. An ordered topological vector space in which order-bounded sets are bounded but the cone is not normal.

Let $E = \phi$ be the space of finite sequences equipped

with supremum norm and ordered by the cone P_s as defined in ≠ 3. As shown there, P_s is not normal. Now,

let $a = (a_1, \ldots, a_k) \varepsilon P_s$, and

$A_r = a_1 + \ldots + a_r$, $(1 \leq r \leq s)$,

$A = \max A_r$.

If $x = (x_n)$ and $0 \leq x \leq a$, then

$$0 \leq \xi_r \leq A$$

for all $r \varepsilon N$, where

$$\xi_r = x_1 + \ldots + x_r .$$

Hence,

$$|x_r| = |\xi_r - \xi_{r-1}| \leq A$$

for all r, that is, $||x|| \leq A$.

5. A cone in a topological vector space, which has no interior points.

(i) The positive cone C defined by

$$C = \{x = (x_n) \varepsilon \ell^1 \ ; \ x_n \geq 0 \text{ for all } n\}$$

in ℓ^1 which is equipped with the usual norm, has no interior points. For, if it had one, it would be an order-unit in ℓ^1;but ℓ^1 does not have order-units.

(ii) Similarly the positive cone in c_0 does not have interior points.

6. An element of a cone in a vector space which is an interior point for one topology but not for another topology.

e_1 is an interior point of the cone P_s(See # 3) in ℓ^1 with respect to the usual norm. However, e_1 is not interior point of P_s with respect to the supremum norm p on ℓ^1 : for, if

$$x_n = \frac{1}{n} (e_1 + \ . \ . \ . + e_{n+1}),$$

then

$$p(x_n) = \frac{1}{n}$$

and

$$e_1 - x_n \notin P_s \ .$$

7. A cone in a locally convex space, which is not a b-cone.

Let $E = C_0[0, 1]$ be the space of continuous (real-valued) functions on the closed interval $[0,1]$ that vanish at 0, equipped with the norm

$$||f|| = \sup \{|f(x)| \ ; \ x\varepsilon[0,1]\}$$

and ordered by the cone K of non-negative and convex functions in E. Then K is not a b-cone : To show this, let f in E be defined by

$$f(x) = \sqrt{x}, \ x\varepsilon[0,1]$$

and let

$$0<\varepsilon<\tfrac{1}{2} \ .$$

Then

$$||f - (g-h)||<\varepsilon \ , \ g,h\varepsilon K$$

implies that

$$g(t)>f(t)-\varepsilon, \ t\varepsilon[0,1] \ .$$

In particular,

$g(4\varepsilon^2) > \varepsilon$;

hence, since g is convex and g(0) = 0, it follows that

$$g(1) = \frac{g(1) - g(0)}{1 - 0} \geq \frac{g(4\varepsilon^2) - g(0)}{4\varepsilon^2 - 0} > \frac{1}{4\varepsilon}$$

and so

$$||g|| \geq \frac{1}{4\varepsilon} \ .$$

Hence K is not a b-cone in E.

8. A base of a cone in a topological vector space, which is not closed.

Consider E = ϕ, the space of finite (real) sequences equipped with supremum norm and ordered by the cone

$$K = \{x = (x_n) \varepsilon E \ ; \ x_n \geq 0 \text{ for all } n\} \ .$$

Then the set

$$B = \{x = (x_n) \varepsilon K \ ; \ \Sigma x_n = 1\}$$

is a base for B. It is not closed, since

$$x^{(n)} = \frac{e_1 + \ . \ . \ . + e_n}{n} \ \varepsilon B$$

and

$$x^{(n)} \to 0 \ .$$

9. An ordered normed space which is not an order-unit normed space though its dual is a base normed space.

The space c_0 of sequences converging to 0, equipped with supremum norm, is a normed space ordered by the positive cone of non-negative sequences. It is not an order-unit-normed space since it does not have an order-unit.

Since the open unit ball $B = \{x \in c_0 \; ; \; ||x|| < 1\}$ is solid and directed upwards, c_0 is an approximate order-unit-normed space by Theorem 7 and hence its dual ℓ^1 is a base normed space by Proposition 3.

10. An ordered topological vector space which is complete but not order-complete.

(i) The Banach lattice $C[0,1]$ of continuous functions on the closed interval $[0,1]$, equipped with the supremum norm and ordered by the positive cone, is clearly complete. But it is not order-complete.

(ii) The Banach lattice c of convergent sequences is complete but not order-complete.

11. An ordered topological vector space which is order-complete but not complete.

The normed vector lattice ϕ of finite sequences, equipped with the supremum norm and ordered by the positive cone, is order-complete but not complete.

12. An ordered topological vector space which is complete and order-complete but not boundedly order-complete.

The Banach lattice c_0 of sequences converging to 0, equipped with the supremum norm and ordered by the positive cone, is order-complete. Clearly it is complete, Now let

$$e_n = (1, 1, \ldots , 1,0,0, \ldots) \in c_0$$

where 1 appears in n places, and let

$$B = \{e_n \; ; \; n \in \mathbb{N}\} .$$

Then B is increasing and norm-bounded. But, clearly, B does not have a supremum in c_0 . Hence c_0 is not boundedly order-complete.

13. An order-continuous linear functional on an ordered topological vector space, which is not continuous.

Consider the normed vector lattice ϕ as defined in # 11. The topological dual of ϕ can be identified with ℓ^1. However every sequence $u = (u_n)$ of real numbers determines an order-continuous linear functional f_u on ϕ defined by

$$f_u(x) = \sum_{n=1}^{\infty} x_n u_n \ ; \ x = (x_n) \varepsilon \phi \ .$$

(See B[26], page 56). Consequently there exist order-continuous linear functionals on ϕ that are not norm-continuous.

14. A continuous linear operator on an ordered topological vector space, which is not sequentially order-continuous.

Consider the space

$$\ell^2 = \{x = (x_n) \ ; \ \sum_{n=1}^{\infty} |x_n|^2 < \infty, \ x_n \varepsilon \mathbb{R}\}$$

equipped with the norm

$$||x|| = (\sum_{n=1}^{\infty} |x_n|^2)^{\frac{1}{2}} \ , \ x = (x_n) \varepsilon \ell^2,$$

and ordered by the positive cone

$$K = \{x = (x_n) \varepsilon \ell^2 \ ; \ x_n \geq 0 \text{ for all } n\} \ .$$

Consider the infinite matrix $T = (a_{mn})$ defined by

$$a_{mn} = \begin{cases} \dfrac{1}{m-n} & \text{if } m \neq n \\ 0 & \text{if } m = n \ . \end{cases}$$

Then T is a matrix transformation on ℓ^2 into itself (B[12], page 236). T is continuous for the usual norm on ℓ^2. We show that T is not order-bounded which implies that it is not sequentially order-complete, because the vector space of order-bounded linear operators on ℓ^2 is equal to that of sequentially order-continuous linear operators on ℓ^2 (B[26], page 52,53). If $x = (x_m)$ and $y = (y_m)$ are the elements of ℓ^2 defined by

$$x_m = y_m = \frac{1}{\sqrt{m}\,\log m} \quad (m>1),$$

and

$$x_1 = y_1 = y_2 ,$$

then

$$(*) \quad \sum_{m=1}^{\infty} \sum_{n=1}^{\infty} |a_{mn}| x_m \, y_n = +\infty .$$

Define $z = (z_m) \in \ell^2$ by

$$z_m = \sum_{n=1}^{\infty} |a_{mn}| \, y_n$$

and set

$$u_n^{(m)} = (\operatorname{sign} a_{mn}) \, y_n .$$

Then

$$u^{(m)} = (u_n^{(m)}) \in [-y,y]$$

for all m, and

$$z_m = \sum_{n=1}^{\infty} a_{mn} \, u_n^{(m)} = (T\,u^{(m)})_m .$$

Suppose T is order-bounded. Then there is a $u \geq 0$ such that

$$Tu^{(m)} \in [-u,u]$$

for all positive integers m. It follows that

$$- < u,v > \leq < Tu^{(m)}, v > \leq <u,v>$$

for all $v \geq 0$ in ℓ^2; in particular, if we take v to be the unit vector $e^{(m)}$, we obtain the relation

$$z_m = |z_m| = |(Tu^{(m)})_m| \leq u_m$$

for each m. But then

$$\sum_{m=1}^{\infty} \sum_{n=1}^{\infty} |a_{mn}| y_n x_m = \sum_{m=1}^{\infty} z_m x_m \leq \sum_{m=1}^{\infty} u_m a_m < \infty$$

since $x,u \in \ell^2$. This contradicts (*).

Hence T is not order-bounded.

15. A positive linear functional on an ordered topological vector space, which is not continuous.

Consider the normed vector lattice ϕ as defined in # 11. If $x = (x_n) \in \phi$, the linear functional f defined by $f(x) = \Sigma x_n$ is positive but not continuous.

16. An ordered topological vector space on which there exist no non-zero positive linear functionals.

Consider the complete metrizable topological vector lattice $L^p[a,b]$, $a,b \in \mathbb{R}$, $0<p<1$, as defined in chapter 1, # 3(i). As we have seen in Chapter 1, # 6, there exist no non-zero continuous linear functionals on $L^p[a,b]$. But every positive linear functional on a complete metrizable topological vector lattice is continuous (B[26], page 88).

Hence there exist no non-zero positive linear functionals on $L^p[a,b]$, $0<p<1$.

17. A topological vector lattice which has no non-zero real lattice homomorphisms.

Consider the Banach lattice $E = L^1[0,1]$ with the usual norm and order. Every positive linear functional on E is continuous (B[26], page 88). Let u be a real lattice homomorphism and $u(f) = 1$ for some f with $||f|| = 1$. We can write f in the form $g_1 + g_1'$ where

$$||g_1|| = ||g_1'|| = \tfrac{1}{2}$$

and

$$\inf\{g_1 , g_1'\} = 0 .$$

We have $u(f_1) = 1$ for $f_1 = g_1$ or g_1'. Repeating the construction, we obtain a sequence of elements $\{f_n\}$ such that

$$||f_n|| = 2^{-n} , \; u(f_n) = 1 .$$

contradicting the continuity of f.

18. A topological vector space with lattice ordering in which the map $x \to x^+$ is continuous for all x but not uniformly continuous.

Consider c_0 with the usual norm and the order given by the cone P_s as defined in # 3. As shown there, P_s is not normal. Hence the map $x \to x^+$ is not uniformly continuous. Now we show that it is continuous at any point $x = x_n$ in c_0. Take $\varepsilon > 0$. There exists N such that $|x_r| \leq \varepsilon$ for $r > N$. Take $y = (y_n)$ in c_0 such that

$$||y - x|| \leq \frac{\varepsilon}{N} .$$

Let

$$X_r = x_1 + \ldots + x_r ,$$

$$Y_r = y_1 + \ldots + y_r, \ (r \geq 1)$$

$$X_0 = 0, \ Y_0 = 0 .$$

Then

$$x^+ = (\alpha_r) \ , \ y^+ = (\beta_r)$$

where

$$\alpha_r = X_r^+ - X_{r-1}^+ ,$$

$$\beta_r = Y_r^+ - Y_{r-1}^+ , \ (r \geq 1).$$

For $r \leq N$, we have

$$|Y_r^+ - X_r^+| \leq |Y_r - X_r| \leq \sum_{s=1}^{r} |y_s - x_s| \leq \varepsilon$$

and

$$|\beta_r - \alpha_r| \leq 2\varepsilon \ .$$

For $r > N$,

$$|\alpha_r| \leq |x_r| \leq \varepsilon$$

and

$$|\beta_r| \leq |y_r| \leq \varepsilon + ||y - x|| \leq 2\varepsilon$$

and so

$$|\beta_r - \alpha_r| \leq 3\varepsilon \ .$$

Hence,

$$||y^+ - x^+|| \leq 3\varepsilon.$$

19. An ordered locally convex space with a positive weakly convergent sequence which is not convergent.

Consider the normed space ℓ^∞ of bounded sequences, equipped with the supremum norm, and ordered by the cone P_s as defined in # 3. Let e_n denote the element in ℓ^∞ having 1 in the nth coordinate and zeros elsewhere. Then the sequence $\{e_n\}$ is a decreasing sequence which converges to 0 in the weak topology but not in the norm topology.

20. An M-space which is not normable.

The space of all real-valued functionals on an infinite set X, equipped with the topology of pointwise convergence is an M-space which is not normable.

21. A Pseudo-M-space which is not an M-space.

Consider the normed space $E = \ell^\infty$ or c_0, equipped with the supremum norm and ordered by the cone P_s as defined in # 3. Given $x = (x_n)$ and $y = (y_n)$, let

$$X_n = x_1 + \ldots + x_n ,$$

$$Y_n = y_1 + \ldots + y_n ,$$

$$z_n = (X_n \vee Y_n) - (X_{n-1} \vee Y_{n-1}),\ n \geq 2$$

and

$$z_1 = X_1 \vee Y_1 .$$

It is easily seen that

$$|z_1| \leq |x_n| \vee |y_n|$$

for all n so that

$$||z|| \leq ||x|| \vee ||y||,\ z = (z_n).$$

Also

$$z_1 + \ldots + z_n = X_n \vee Y_n$$

so that $z = x \vee y$. Thus, E is a pseudo-M-space. Since P_s is not normal as shown in # 3, E is not an M-space.

22. A topological vector lattice which is not a pseudo-M-space.

The Banach lattice ℓ^1 with the usual norm and order is not a pseudo-M-space, because any sublattice containing the unit ball is unbounded.

23. The topology of a bornological locally convex lattice which is not an order bound topology.

Consider the Banach lattice $E = C[0,1]$ as defined in # 10(i). Let F be the subspace of E consisting of all elements f in E which vanish in a neighbourhood (depending on f) of $t = 0$, K the positive cone in F. Then F is a normed vector lattice under the induced topology and so a bornological space. Let

$$V = \{f \in F \ ; \ |f(\tfrac{1}{n})| \leq \tfrac{1}{n} \ , \ n = 1,2, \ . \ . \ .\} \ .$$

Clearly V is closed, convex and solid.

If $f \in F$, let a_f, $0 < a_f < 1$, be such that $f(t) = 0$ for all $t \in |0, a_f|$. If

$$\lambda = a_f \ ||f||^{-1},$$

then $\lambda x \in V$ and so V is absorbing. We now show that V is not a neighbourhood of 0. For this, it is sufficient to show that for any natural number $r \geq 1$, there is an $f_r \in F$ with $||f_r|| \leq \frac{1}{r}$ such that $x_r \notin V$; it follows, then, that 0 is not an interior point of V and so V is not a neighbourhood of 0. Consider $A = [0, \frac{1}{4r}]$ and $B = [\frac{1}{r+1} \ , \ 1]$ in $[0,1]$.

By Urysohn's lemma, there is a continuous real-valued function f_r on $[0,1]$ with range in $[0, \frac{1}{r}]$ such that

$$f_r(t) = \begin{cases} 0 & \text{if } t \in A \\ \frac{1}{r} & \text{if } t \in B \end{cases} .$$

Clearly $f_r \in F$ and

$$\|f_r\| \leq \frac{1}{r} .$$

On the other hand, since

$$f_r \left(\frac{1}{r+1} \right) = \frac{1}{r} > \frac{1}{r+1} ,$$

it follows that $f_r \notin V$. Thus, V is not a neighbourhood of 0. Since a solid set is absorbing iff it is order-bornivorous, we conclude that the topology on F is not an order bound topology.

24. An order-quasibarrelled vector lattice which is not barrelled.

Cf. # 26.

25. A C.O.Q. vector lattice which is not order-quasi-barrelled.

The locally convex space C(W) as defined in Chapter 3, # 41(ii), ordered by the positive cone

$$K = \{f \in C(W) \; ; \; f(t) \geq 0 \text{ for all } t \in W\}$$

is a locally convex lattice. We have seen in Chapter 3, # 41(ii), that C(W) is countably barrelled; hence it is a C.O.Q. vector lattice. But it is not order-quasibarrelled because it is not quasibarrelled as shown in Chapter 3, # 41(ii).

26. A C.O.Q. vector lattice which is not countably barrelled.

Consider the Banach lattice $E = \ell^\infty$ of bounded sequences, equipped with the supremum norm and the usual pointwise ordering. Let e_n be the sequence in E having 1 in the nth coordinate and zeros elsewhere, F_0 the subspace of E generated by the e_n (F_0 consists of all finite sequences), e the sequence having 1 in every coordinate and

$$F = \{x + \lambda e \ ; \ x \varepsilon F_0 \ , \ \lambda \varepsilon \mathbb{R}\} .$$

Then with the induced norm and order, F is an order-unit-normed vector lattice with order-unit e. It is easy to see that an order-unit-norm topology is an order bound topology and that each vector lattice with an order bound topology is order-quasibarrelled. Thus F is order-quasibarrelled. Now, define f_n on F by

$$f_n(x + \lambda e) = x_n \ , \ x = (x_n) \varepsilon F_0 \ .$$

Then $\{f_n\}$ is a pointwise bounded sequence of continuous linear functionals but $\{f_n\}$ is not equicontinuous. Hence F is not countably barrelled.

27. A countably quasibarrelled locally convex lattice which is not a C.O.Q. vector lattice.

The normed vector lattice F as defined in ≠ 23 is quasibarrelled and hence countably quasibarrelled. But it is not a C.O.Q. vector lattice. For, if it were, then, being metrizable, it would be order-quasibarrelled, because a quasibarrelled C.O.Q. vector lattice is order-quasibarrelled (B[15] or P[28]). But it is not an order-quasi-barrelled vector lattice.

28. An order-quasibarrelled (hence a C.O.Q.) vector lattice which is not an order-(DF)-vector lattice.

Consider the Banach lattice ℓ^1 with the usual norm and pointwise ordering. Clearly ℓ^1 is an order-quasi-barrelled vector lattice. We show that it does not have a fundamental sequence of order-bounded sets. Suppose it has. Then $\sigma_s(\ell^\infty,\ell^1)$ is metrizable. Since $(\ell^\infty, \sigma_s(\ell^\infty,\ell^1))$ is complete, it follows that $(\ell^\infty, \sigma_s(\ell^\infty, \ell^1))$ is complete and metrizable and hence $\sigma_s(\ell^\infty, \ell^1)$ coincides with the supremum norm topology of ℓ^∞, because the two topologies are comparable. Thus, $\sigma_s(\ell^\infty, \ell^1)$ is normable. Also the positive cone in ℓ^1 is generating and $\sigma_s(\ell^1, \ell^\infty)$-closed. But then ℓ^1 has an order-unit (B[26]), which is not true.

29. An O.Q.U. vector lattice which is not ultrabarrelled.

The normed vector lattice F, as defined in # 26, is not barrelled and hence not ultrabarrelled, because a locally convex space which is ultrabarrelled is barrelled. But it is an O.Q.U. vector lattice because it is an order-unit-normed space.

30. A quasi-ultrabarrelled topological vector lattice which is not an O.Q.U. vector lattice.

The normed vector lattice F, as defined in # 23, is not an O.Q.U. vector lattice, because it is not order-quasibarrelled as can be seen in # 23. However, since a normed space is quasi-ultrabarrelled, so is F.

31. An order-quasibarrelled vector lattice which is not an

O.Q.U. vector lattice.

Cf. # 34.

32. A countably O.Q.U. vector lattice which is not countably ultrabarrelled.

The order-unit-normed vector lattice F, as in # 26, is an O.Q.U. vector lattice, as shown in # 29, and hence countably O.Q.U. vector lattice. But F is not countably barrelled, as shown in # 26, and hence not countably ultrabarrelled either. Note that a locally convex space which is countably ultrabarrelled is countably barrelled.

33. A countably quasi-ultrabarrelled topological vector lattice which is not a countably O.Q.U. vector lattice.

The normed vector lattice F, as defined in # 23, is quasi-ultrabarrelled, being a normed space, and hence countably quasi-ultrabarrelled. Since it is not a C.O.Q. vector lattice, as shown in # 27, it is not a countably O.Q.U. vector lattice either. Note that a locally convex space which is countably O.Q.U. space is a C.O.Q. space.

34. A C.O.Q. vector lattice which is not a countably O.Q.U. vector lattice.

The space $(\ell^{\frac{1}{2}}, u)$, as defined in Chapter 4, # 6, is infact a complete and metrizable topological vector lattice for the order induced by the cone

$$K = \{x = (x_i) \varepsilon \ell^{\frac{1}{2}} \ ; \ x_i \geq 0 \text{ for all } i\} \ .$$

As we have seen there, $(\ell^{\frac{1}{2}}, u^{00})$ is barrelled and hence order-quasibarrelled which, in turn, implies that it is a C.O.Q. vector lattice. Now, for each $x = (x_i) \in \ell^{\frac{1}{2}}$, define

$$t_n(x) = (x_1, \ . \ . \ . \ , x_n \ , 0,0, \ . \ . \ . \ ).$$

Then each t_n is a continuous lattice homomorphism of $(\ell^{\frac{1}{2}}, K, u^{00})$ into $(\ell^{\frac{1}{2}}, K, u)$ and for each $x \in \ell^{\frac{1}{2}}$, $f_n(x) \to x$ under u. But $\{t_n\}$ is not equicontinuous as observed in Chapter 4, # 16, and hence $(\ell^{\frac{1}{2}}, K, u^{00})$ is not a countably O.Q.U. vector lattice by Theorem 12.

CHAPTER 6

HEREDITARY PROPERTIES

Introduction

Let E be a vector space. A nonempty subset M of E is called a (vector) subspace of E if $M+M \subset M$ and $\lambda M \subset M$. If E is a topological vector space, then a subspace M of E is a topological vector space under the induced topology.

Let M be a subspace of a vector space E. Consider

$$G = \{x + M \; ; \; x \in E \text{ such that } x+M = y+M \text{ iff } x - y \in M\}$$

In G, define the addition operation as

$$(x+M) + (y+M) = x+y+M$$

and scalar multiplication by

$$\lambda(x+M) = \lambda x+M .$$

Then G becomes a vector space called quotient vector space which is denoted by E/M. Now let E be a topological vector space, M a subspace of E and ϕ the map of E onto E/M defined by $\phi(x) = x+M$. ϕ is called the canonical or quotient map. The quotient topology on E/M is defined to be the finest topology on E/M for which ϕ is continuous. E/M, equipped with this topology, is called a quotient space.

Let $\{E_\alpha\}$ be a family of vector spaces and $E = \prod_\alpha E_\alpha$ the cartesian product. E becomes a vector space with pointwise addition and scalar multiplication. If $\{E_\alpha\}$ is a family of topological vector spaces, then $E = \prod_\alpha E_\alpha$ is also a topological vector space with the basis of neighbourhoods $U = \prod_\alpha W_\alpha$, when finitely many W_α are circled neighbourhoods U_α of 0 in E_α and all the other $W_\alpha = E_\alpha$.

1. A closed subspace of a reflexive space, which is not reflexive.

Let w denote the space of all sequences $x = (x_n)$ and ϕ the space of finite sequences. We denote by ϕw the (topological) countable direct sum of copies of w, because all the vectors of this direct sum are obtained when in each vector $(x_1, \ldots x_n, 0,0, \ldots)$ of ϕ the non-zero x_i are replaced by arbitrary non-zero vectors u_i of w and the zeros are replaced by the zero vector 0 of w. Similarly we denote by $w\phi$ the (topological) countable product of copies of ϕ. Let $E = \phi w + w\phi$. Then its dual is $E' = w\phi + \phi w$. Since the locally convex direct sum of Montel spaces is a Montel space, E' is a Montel space and hence reflexive. Let

$$M = \{(x,x) \ ; \ x \varepsilon \phi w \cap w\phi = \phi\}.$$

Then $N = M^{\perp}$ consists of all $(x,-x)$, $x \varepsilon \phi$, in $E' = w\phi + \phi w$. N is a closed subspace of E', which is not reflexive in the relative topology.

2. A closed subspace of a bornological space, which is not bornological.

Clearly $E' = w\phi + \phi w$ is bornological (because the locally convex direct sum of bornological spaces is bornological).

The closed subspace N, as defined in # 1, is not bornological; if it were, then, being complete, it would be barrelled. But then, being semi-reflexive, it would be reflexive which is not true.

3. An infinite countable codimensional subspace of a

bornological space, which is not quasibarrelled.

Cf. Valdivia, M. : Some examples on quasibarrelled spaces, Ann. Inst. Fourier, Grenoble, 22, 2(1972), 21-26.

4. A closed subspace of a barrelled space which is not countably quasibarrelled.

Since any Hausdorff locally convex space is a closed subspace of some barrelled space (P[49]), it is now sufficient to give an example of a Hausdorff locally convex space which is not countably quasibarrelled. Let u be the supremum norm on the space c_0 of sequences convergent to 0, and v the associated weak topology on c_0. For each n, let

$$g_n : (E,v) \to (E,u)$$

be defined by

$$g_n(x) = (x_1, x_2, \ldots, x_n, 0, 0, \ldots).$$

Then $\{g_n\}$ is a sequence of continuous linear maps from (E,v) into (E,u) such that, for each x in E, $g_n(x)$ converges to x in (E,u). Moreover, $\{g_n\}$ is uniformly bounded on bounded sets, for if B is the unit ball in (E,u), the union over n of $g_n(B)$ is contained in B. But $\{g_n\}$ is not equi-continuous since v is strictly coarser than u and so (E,v) is not countably quasibarrelled.

5. A dense uncountable dimensional subspace of a barrelled space, which is not barrelled.

Let s be the space of real sequences with the product topology. Let m be the dense subspace of s consisting of the bounded sequences. Then m is not barrelled in the relative topology because the barrel $\{x = (x_n) \in m$;

$|x_n| \leq 1$, $n \in \mathbb{N}\}$ is not a neighbourhood of 0. Of course m has uncountable dimension, because a countable codimensional subspace of a barrelled space is barelled.

6. A closed subspace of a (DF)-space, which is not a (DF)-space.

Let E be the vector space (over $\mathbb{R}$) of all double sequences $x = (x_{ij})$ such that for each $n \in \mathbb{N}$,

$$p_n(x) = \sum_{i,j} a_{ij}^{(n)} |x_{ij}| < \infty$$

where

$$a_{ij}^{(n)} = \begin{cases} j^n & \text{for } i < n \\ i^n & \text{for } i \geq n \end{cases} .$$

Equipped with the topology generated by the semi-norms $\{p_n\}$, E is a Fréchet Montel space. The dual E' of E can be identified with the space of all double sequences $u = (u_{ij})$ such that $|u_{ij}| \leq c\, a_{ij}^{(n)}$ for all i, j and suitable $c>0$, $n \in \mathbb{N}$ (the canonical bilinear functional being $(x, u) \to \langle x, u \rangle = \sum_{i,j} x_{ij} u_{ij}$). Each $x \in E$ defines a summable family $\{x_{ij} ; (i,j) \in \mathbb{N} \times \mathbb{N}\}$. Let T be the linear map from E which sends each $x = (x_{ij}) \in E$ to the vector

$$Tx = \left(\sum_{i=1}^{\infty} x_{i1} , \sum_{i=1}^{\infty} x_{i2}, \ldots \right).$$

It follows that

$$\sum_{j=1}^{\infty} \left| \sum_{i=1}^{\infty} x_{ij} \right| \leq \sum\sum |x_{ij}| < \infty$$

and so $Tx \in \ell^1$. T is a continuous linear map of E onto a dense subspace of ℓ^1. The adjoint T' maps each element $\xi \in \ell^\infty$ to the element $T'\xi \in E'$. The subspace of all $T'\xi$ is weakly closed in E'. T' is thus a weakly continuous one-one linear map from ℓ^∞ into E' with a weakly closed image space. It follows that T is a topological homomorphism of E onto ℓ^1 (Schaefer [31], page 160).

Thus E/N, $N = \{x \varepsilon E ; T(x) = 0\}$, is isomorphic with ℓ^1. Let ϕ be the canonical map of E onto E/N . Every closed bounded subset B of E is compact, and so its image $\phi(B)$ is compact. But there exist bounded sets, which are not compact in $E/N \cong \ell^1$; not every bounded subset of E/N is contained in the closure of the image $\phi(B)$ of a bounded set B. So, the $\mathfrak{G}$-topology on the closed subspace $H=T'(\ell^\infty)$, where $\mathfrak{G}$ is the family of all relatively compact subsets of E/N, is strictly finer than $\beta(E',E)$. H is not quasi-barrelled for this topology. Since the strong dual of a Fréchet space is a (DF)-space, $(E',\beta(E',E))$ is a (DF)-space. But the closed subspace H is not a (DF)-space, because a separable (DF)-space is quasibarrelled. (We note that H is separable).

7. An infinite countable codimensional subspace of a quasi-barrelled (DF) space, which is not a (DF) space.

Cf. Valdivia's paper mentioned in # 3.

8. A closed subspace of a hyperbarrelled space, which is not hyperbarrelled.

Since a Hausdorff locally semiconvex space is a closed subspace of some Hausdorff hyperbarrelled space, (Cf. P[34]), # 3 is an example of a closed subspace of a hyperbarrelled space which is not γ-quasi-hyperbarrelled, a fortiori, not hyperbarrelled.

9. A closed subspace of an ultrabarrelled space, which is not countably quasi-ultrabarrelled.

The space $(E, \tau(E,E^*))$ as defined in Chapter 4, # 16, is not countably quasi-ultrabarrelled, a fortiori, not ultra-barrelled. Since $(E, \tau(E,E^*))$ is Hausdorff and complete, it can be embedded as a closed subspace of a product G of Banach spaces. Since G is of the second category in itself, it is ultrabarrelled while the closed subspace $(E, \tau(E,E^*))$ is not.

10. A lattice ideal in an order-quasi-barrelled vector

lattice, which is not order-quasibarrelled.

The subspace F of E as defined in Chapter 5, # 23 is infact a lattice ideal in E. But F is not order-quasibarrelled.

11. A complete locally convex space whose quotient is not sequentially complete.

The locally convex space $E = \phi w + w\phi$, as defined in # 1, is complete. We show that the quotient space E/M, where M is the closed subspace as defined in # 1, is not sequentially complete. Given a sequence $x^{(n)} = (\eta^{(n)}, z^{(n)})$ in E, the cosets $\dot{x}^{(n)}$ with respect to M form a Cauchy sequence if the sequence is almost constant in each coordinate. In particular, if we take the sequence

$$\dot{x}^{(n)} = \left(\sum_{i,k=1}^{n} e_{ik} \, , \, -\sum_{i,k=1}^{n} e_{ik} \right)$$

then $\dot{x}^{(n)}$ is a Cauchy sequence. But this has no limit in E/M; for, there is no $x = (\eta\ , z) \varepsilon E$ for which $\dot{x} - \dot{x}^{(n)}$ converges to $\dot{0}$.

12. A quotient of a Montel space, which is not semi-reflexive.

The space $E = \phi w + w\phi$ is a Montel space. The quotient space E/M as in # 11 is not semi-reflexive since it is not weakly sequentially complete. Note that a locally convex space is semi-reflexive iff it is weakly quasi-complete.

13. A quotient of a Fréchet Montel space, which is not reflexive.

The locally convex space E, as defined in # 6, is a Fréchet Montel space. But the quotient space E/N, as defined there, is not reflexive because E/N is topologically isomorphic to ℓ^1 and ℓ^1 is not reflexive.

14. A product of B-complete spaces which is not B-complete.

Let E be a product of Banach spaces which are, of course, B-complete. Suppose E is B-complete. Then, since every complete locally convex space is a closed subspace of a product of Banach spaces and since every closed subspace of a B-complete space is B-complete, it follows that every complete locally convex space is B-complete which is not true as shown in Chapter 2, # 5.

15. An arbitrary direct sum of B-complete spaces, which is not B-complete.

Take $E_W = \sum_{\alpha \in I} \mathbb{R}_\alpha$ in # 5 of Chapter 2, where $\mathbb{R}_\alpha$ is a copy of the real line. A Fréchet space is B-complete and so each $\mathbb{R}_\alpha$ is B-complete. However, E_W is not B-complete as shown in Chapter 2, # 5.

CHAPTER 7

TOPOLOGICAL BASES

Introduction

In this chapter we present some counter-examples in topological bases.

One of the celebrated queries of the subject was the following:

Does there exist a basis for every separable Banach space? This query appeared in Banach's Monograph B [1] and remained unsolved till the appearance of Enflo's paper P [21] where the query has been answered in the negative by a counter-example which is the first counter-example presented in this chapter. We, first, collect some definitions and results pertaining to the subject matter of this chapter.

Let E be an infinite-dimensional Banach space. Let the series $\sum_{i=1}^{\infty} x_i$ be convergent in E in which case we write

$x = \sum_{i=1}^{\infty} x_i$.

$\sum_{i=1}^{\infty} x_i$ is called unconditionally convergent if for every permutation r of the integers the series $\sum_i x_{r(i)}$ converges in E.

THEOREM 1. Let E be a Banach space and $\sum_i x_i$ a convergent series in E. The following statements are equivalent.

(a) $\sum_i x_i$ is unconditionally convergent.

(b) $\sum_i x_i$ is unordered convergent (that is,

$\lim_{\sigma \in \Phi} \sum_{i \in \sigma} x_i = x$, where Φ is the set of all finite subsets of $\mathbb{N}$, directed by $\supseteq$).

(c) $\sum_i x_i$ is subseries convergent (that is, if for every increasing sequence $\{n_i\}$ of integers, the series $\sum_i x_{n_i}$ converges to some element of E).

(d) $\sum_i x_i$ is bounded multiplier convergent (that is, if for each bounded sequence $\{\alpha_i\}$ in $\mathbb{K}$, the series $\sum_i \alpha_i x_i$ is convergent to some element of E). (Cf. B [22]).

A series $\sum_{i=1}^{\infty} x_i$ in a Banach space E is said to be absolutely convergent if $\sum_{i=1}^{\infty} ||x_i||$ is convergent.

Every absolutely convergent series is unconditionally convergent. The converse is not true for all infinite-dimensional Banach spaces P[19] .

A (topological) basis for an infinite dimensional Banach space E is a sequence $\{x_i\}$ in E such that to every $x \in E$, there corresponds a unique sequence $\{\alpha_i\}$ in $\mathbb{K}$ for which $x = \lim_n \sum_{i=1}^{n} \alpha_i x_i$ in the strong topology of E. If E is finite-dimensional, it is just the Hamel basis. If the limit is taken in the weak topology of E, then we call $\{x_i\}$ the weak basis for E.

The elements of the sequence $\{\alpha_i\}$ which depend linearly on x are called the coefficient functionals of the basis $\{x_i\}$. The uniqueness of $\{\alpha_i\}$ implies that every element in $\{x_i\}$ is non-zero. A basis which has continuous coefficient functionals is called a Schauder basis. Every basis for a Banach space (even for a Fréchet space) is a Schauder basis.

A basis $\{x_i\}$ for a Banach space E is said to be boundedly complete if for each sequence $\{\alpha_i\}$ in $\mathbb{K}$ with $\sup_n \|\sum_{i\leq n} \alpha_i x_i\| < \infty$, there exists an x in E such that

$x = \lim_n \sum_{i\leq n} \alpha_i x_i$.

Let $\{x_i\}$ and $\{f_i\}$ be sequences in a Banach space E and its dual E' respectively. (x_i, f_i) is called a biorthogonal system for E if $f_i(x_j) = \delta_{ij}$.

THEOREM 2. $\{x_i\}$ is a (weak) basis for E iff there is a sequence $\{f_i\}$ in E' such that (x_i, f_i) is a biorthogonal system for E and $\sum_{i=1}^{n} f_i(x)x_i$ converges strongly (respectively, weakly) to x for each x in E. (Cf. B[22])

COROLLARY 1. (weak basis theorem) $\{x_i\}$ is a basis for the Banach space E iff it is a weak basis for E.

A basis (x_i, f_i) for a Banach space E is said to be unconditional if for all x in E, the series $\sum_{i=1}^{\infty} f_i(x)x_i$ converges unconditionally. It is said to be absolutely convergent if $\sum_{i=1}^{\infty} f_i(x)x_i$ converges absolutely for every x in E. It is called normal if $\|x_i\| = \|f_i\| = 1$, $i = 1,2, \ldots$. It is called E-complete biorthogonal system if the sequence $\{x_i\}$ is complete in E in the sense that the set of all finite linear combinations $\sum_{i=1}^{n} \alpha_i x_i$, $\alpha_i \in \mathbb{K}$, $n = 1,2, \ldots$, is dense in E.

A basis $\{x_i\}$ of a Banach space E is

(i) monotone if

$$\|\sum_{i=1}^{n} \alpha_i x_i\| \leq \|\sum_{i=1}^{n+m} \alpha_i x_i\|$$

for all finite sequences of scalars $\alpha_1, \ldots \alpha_{n+m}$.

(ii) symmetric if

$$\sup_{\sigma \in P} \sup_{\substack{|\beta_i| \leq 1 \\ 1 \leq k < \infty}} \left\| \sum_{i=1}^{k} \beta_i f_i(x) x_{\sigma(i)} \right\| < \infty,$$

where $\{f_i\}$ are the coefficient functionals and P is the set of all permutations of $\mathbb{N}$;

(iii) subsymmetric if it is unconditional and for every increasing sequence of positive integers $\{i_k\}$, the basis $\{x_{i_k}\}$ of the space spanned by the sequence $\{x_{i_k}\}$ is equivalent to the basis $\{x_i\}$;

(iv) Besselian if $\sum_{i=1}^{\infty} |\alpha_i|^2 < \infty$ whenever $\sum_{i=1}^{\infty} \alpha_i x_i$ is convergent; and

(v) Hilbertian if $\sum_{i=1}^{\infty} \alpha_i x_i$ is convergent whenever $\sum_{i=1}^{\infty} |\alpha_i|^2 < \infty$.

If E is a Banach space, a basis $\{f_i\}$ of E' is called retro-basis if, for the coefficient functionals $\{\phi_i\}$ in E", $\{\phi_i\} \subset \pi(E)$ where π is the canonical map of E into E".

Let E be a topological vector space, E' its dual and I any index set. Let $\{x_\lambda\}_{\lambda \in I}$ and $\{f_\lambda\}_{\lambda \in I}$ be families of elements in E and E' respectively. (x_λ, f_λ) is called a biorthogonal system if $f_\lambda(x_\mu) = \delta_{\lambda\mu}$. It is maximal with respect to E if there is no biorthogonal system which contains (x_λ, f_λ) properly. (x_λ, f_λ) is called a generalized basis for E if $f_\lambda(x) = 0$, $\lambda \varepsilon I$, implies that $x = 0$ for all x in E. It is called an extended Markushevich basis for E if $\{x_\lambda\}$ is total in E; in particular, if I is countable, then (x_λ, f_λ) is called a Markushevich basis for E.

As in the case of Banach space, a sequence $\{x_i\}$ in a topological vector space E is a basis for E if for each x in E, there is a unique sequence $\{\alpha_i\}$ in $\mathbb{K}$ such that

$$x = \lim_n \sum_{i \leq n} \alpha_i x_i$$

in the topology of E. Each expansion coefficient α_i defines a linear functional f_i, $f_i(x) = \alpha_i$, on E. These coefficient functionals f_i, however, need not be continuous (see # 21). If the coefficient functionals $\{f_i\}$ are continuous, then $\{x_i\}$ is called a Schauder basis.

Let (x_i, f_i) be a Schauder basis for a topological vector space E. Let

$$s_n(x) = \sum_{i=1}^{n} f_i(x)\, x_i \ , \ x \varepsilon E, \ n = 1,2, \ldots .$$

$\{x_i\}$ is called

(i) an (e)-Schauder basis of E if $\{s_n\}$ is an equicontinuous subset of $\mathcal{L}(E)$, the space of continuous linear maps of E into itself;

(ii) a (b)-Schauder basis of E if $\{s_n\}$ is a bounded subset of $\mathcal{L}(E)$, where $\mathcal{L}(E)$ is equipped with the topology of uniform convergence on bounded subsets of E.

If E is a normed space, then (e)-Schauder bases and

(b)-Schauder bases are equivalent.

For a topological vector space the following implications hold:

Schauder basis
$\Downarrow$
Markuschevich basis
$\Downarrow$
Extended Markushevich basis
$\Downarrow$
Generalized basis
$\Downarrow$
Maximal orthogonal system.

Let E and F be two topological vector spaces. A sequence $\{x_i\}$ in E is said to be similar to a sequence $\{y_i\}$ in F if for all sequences $\{\alpha_i\}$ in $\mathbb{K}$, $\sum_{i=1}^{\infty} \alpha_i x_i$ converges (in E) iff $\sum_{i=1}^{\infty} \alpha_i y_i$ converges (in F).

THEOREM 3. (Isomorphism Theorem) Let E and F be barrelled spaces and (x_i, f_i) and (y_i, g_i) Schauder bases in E and F respectively. Then (x_i, f_i) is similar to (y_i, g_i) iff there is a topological isomorphism $T:E \to F$ such that $Tx_i = y_i$ for all $i \geq 1$. (Cf. B [15])

1. A separable Banach space which has no basis.

Cf. P [21] .

2. A Banach space with a basis, whose dual does not have a basis.

Consider the Banach space $E = \ell^1$. It has a basis, namely $\{\delta_i\}$ where $\delta_i = (\delta_{ij})_{j=1}^{\infty}$, δ_{ij} being the Kronecker delta. But the dual of E is ℓ^{∞} which is not separable and does not have a basis.

3. A Banach space which has no unconditional basis.

Consider the Banach space $E = C[0,1]$ of continuous functions on the closed interval $[0,1]$ with the supremum norm topology. Since E' is not separable, it does not have a basis. Now, since E' is weakly sequentially complete, E has no unconditional basis in view of the following: If (x_n, f_n) is an unconditional basis in a Banach space X and if X' is weakly sequentially complete, then $\{f_n\}$ is an unconditional basis in X'.

4. A Banach space with a basis which is not unconditional.

Consider the Banch space c_0 of sequences converging to 0, equipped with the supremum norm topology. Let

$$x_{ij} = \begin{cases} 1 & \text{if } j \leq i \\ 0 & \text{if } j > i \,. \end{cases}$$

Let $\{x_{ij}\} = x_i$. We show that $\{x_i\}$ is a basis for c_0. Let $\{\delta_{ij}\} = \delta_i$, where δ_{ij} is the Kronecker delta. Then the sequence $\{f_i\}$ in ℓ^1, given by

$$f_i = \delta_i - \delta_{i+1} \,,$$

is biorthogonal to $\{x_i\}$. Since the closure of the linear span of $\{x_i\}$ is c_0, and since

$$\sup_n ||T_n\alpha|| = \sup_n || \sum_{i \leq n} f_i(\alpha)x_i||$$

$$= \sup_n || \sum_{i \leq n} (\alpha_i - \alpha_{i+1})x_i||$$

$$= \sup_n \sup \{|\alpha_j - \alpha_n| \; ; \; j<n \}$$

$$\leq 2 \sup_n |\alpha_n| = 2 ||\alpha|| ,$$

it follows that $\{x_i\}$ is a basis for c_0 in view of the following: If (x_i, f_i) is a biorthogonal system for a Banach space E such that $\sup |f(T_n x)| < \infty$, $x \varepsilon E$, $f \varepsilon E'$, then $\{x_i\}$ is a basis for the closure of the linear span of $\{x_i\}$ and $\{f_i\}$ is a basis for the closure of the linear span of $\{f_i\}$. Now, we show that $\{x_i\}$ is not unconditional. Let

$$\alpha = \frac{(-1)^{i+1}}{i} \varepsilon c_0 .$$

We then take the subseries of all odd terms in the expansion of α : With i as odd integer,

$$\sum_{i=1}^{\infty} f_i(\alpha)x_i = \sum_{i=1}^{\infty} (\alpha_i - \alpha_{i+1}) x_i$$

$$= \{ \sum_{i=2j-1}^{\infty} (-1)^{i+1} \alpha_i \}$$

$$= \{ \sum_{i=2j-1}^{\infty} i^{-1} \} .$$

We conclude that the series expansion for α is not subseries convergent (because the series $1 + \frac{1}{2} + \frac{1}{3} + . . .$ is divergent) and hence not unconditional convergent.

5. A Banach space with an unconditional basis which is not boundedly complete.

Consider the Banach space c_0 . Let $\alpha = \{\alpha_i\} \varepsilon c_0$.

Then $\alpha = \lim_n \sum_{i \leq n} \alpha_i \delta_i$.

In view of the equation $\alpha_i = f_i(\alpha)$ where $f_i \varepsilon c_0' = \ell^1$ is given by $f_i = \delta_i$, and Theorem 2, $\{\delta_i\}$ is a basis for c_0 .

The basis is unconditional, since the definition of supremum norm in c_0 implies that the expansion for α is subseries convergent, and hence unconditionally convergent. Now, the sequence $\alpha = (1,1, . . .)$ in $\mathbb{K}$ is such that

$$\sup_n \, || \sum_{i \leq n} \alpha_i \, \delta_i || = 1 .$$

But $\sum_{i \leq n} \alpha_i \, \delta_i$ is not convergent in c_0 .

6. A Banach space with a basis which is not absolutely convergent.

Consider the Banach space $E = \ell^p$, $1<p<\infty$. The biorthogonal sequence to $\{\delta_i\}$ is $\{f_i\}$, where $f_i \varepsilon \ell_p' = \ell_q$ $(\frac{1}{p} + \frac{1}{q} = 1)$ is given by $f_i = \delta_i$. Since for $\alpha = \{\alpha_i\} \varepsilon \ell^p$, $f_i(\alpha) = \alpha_i$, $\sum_{i \leq n} f_i(\alpha)\delta_i$ converges to α, and so $\{\delta_i\}$ is a basis in ℓ^p . Now, let $\{n_i\}$ be any increasing sequence of integers. Since ℓ^p is complete and

$$|| \sum_{n_i \varepsilon (\mathbb{N})} f_{n_i}(\alpha) \, \delta_{n_i} || = \Big(\sum_{n_i \varepsilon (\mathbb{N})} |\alpha_{n_i}|^p \Big)^{1/p}$$

$$\leq \Big(\sum_{i \varepsilon (\mathbb{N})} |\alpha_i|^p \Big)^{1/p}$$

for each α in ℓ^p and each finite subset $(\mathbb{N})$ of integers, the expansion for α is subseries convergent. Thus, $\{\delta_i\}$ is an unconditional basis by Theorem 1. To show that $\{\delta_i\}$ is not absolutely convergent, let p' be such that $1<p'<p$. Then

$$\alpha = \{ i^{-p'/p} \} \varepsilon \ell^p .$$

But

$$\sum_{i\leq n} ||f_i(\alpha)\ \delta_i|| = \sum_{i\leq n} |\alpha_i| = \sum_{i\leq n} i^{-p'/p}$$

$$\geq \int_1^n t^{-p'/p}\, dt$$

$$= \left(1-\frac{p'}{p}\right)^{-1} \left(n^{1-p'/p}-1\right)$$

which diverges as $n \to \infty$.

7. A Banach space with a basis which is not a normal basis.

Consider the Banach space $E = C\,[0,1]$ as defined in # 3. The sequence

$$x_0(t) = 1\ ,\ x_1(t) = t\ ,$$

$$x_{2^k+\ell}(t) = \begin{cases} 0 & \text{if } t \notin \left(\frac{2\ell-2}{2^{k+1}}, \frac{2\ell}{2^{k+1}}\right), \\ 1 & \text{if } t = \frac{2\ell-1}{2^{k+1}}, \\ \text{linear in} & \left[\frac{2\ell-2}{2^{k+1}}, \frac{2\ell-1}{2^{k+1}}\right] \text{ and} \\ & \left[\frac{2\ell-1}{2^{k+1}}, \frac{2\ell}{k+1}\right] \end{cases}$$

($\ell = 1,2,\ \ldots, 2^k$; $k = 0,1,2,\ \ldots$) constitutes a basis for $C\,[0,1]$. But the sequence of coefficient functionals

$$f_0(x) = x(0),\ f_1(s) = x(1) - x(0)\ ,$$

$$f_{2^k+1}(x) = x\left(\frac{2\ell-1}{2^{k+1}}\right) - \tfrac{1}{2}\, x\left(\frac{2\ell-2}{2^{k+1}}\right) - \tfrac{1}{2} x\left(\frac{2\ell}{2^{k+1}}\right)$$

($x \varepsilon C\,[0,1]$; $\ell = 1,2,\ \ldots,\ 2^k$; $k = 0,1,2,\ \ldots$)

is such that

$||f_0|| = 1,\ ||f_n|| = 2, (n = 1,2, \dots)$.

8. A Banach space whose dual space has a normal basis which is not a retro-basis.

Let E be the real vector space of all real sequences $x = (x_n)$ with

$$\lim_n x_n = 0$$

and

$$||x|| = \sup \{ \sum_{i=1}^{n} (x_{k_{2i-1}} - x_{k_{2i}})^2 + (x_{k_{2n+1}})^2 \}^{\frac{1}{2}} < \infty$$

where the supremum is taken over all positive integers n and finite increasing sequences of positive integers $k_1, \dots, k_{2n+1}$. E is a Banach space. Define

$$\delta_i = (\delta_{ij})_{j=1}^{\infty} .$$

Then $\{\delta_i\}$ is a basis in E. Let $\{f_i\}$ be the corresponding coefficient functionals. Define a sequence $\{g_i\}$ in E' by

$$g_1 = f_1 ,\ g_n = f_{n-1} - f_n \ (n = 2,3, \dots).$$

Then $\{g_i\}$ is a normal basis for E', but not a retrobasis for E'. (For a proof, the interested reader is referred to B [33], page 283).

9. A Banach space with a Besselian basis which is not a Hilbertian basis.

The unit vector basis $\delta_i = (\delta_{ij})_{j=1}^{\infty}$ of the Banach space ℓ^1 is a Besselian basis which is not a Hilbertian basis.

10. A Banach space with a Hilbertian basis which is not

a Besselian basis.

The unit vector basis $\delta_i = (\delta_{ij})$ of the Banach space c_0 is Hilbertian but not Besselian.

11. A Banach space with a basis which is not a monotonic basis.

Consider the Banach space $E = C[0,1]$ as defined in # 3.

Define

$$x_i(t) = t_i , \; i = 1,2, \ldots , \; t \in [0,1] .$$

We can select an infinite subsequence $\{x_{i_k}\}$ of $\{x_i\}$, which is a basic sequence in E. Then the (closed) subspace E_0 spanned by the subsequence $\{x_{i_k}\}$ has no monotonic basis. (For a proof, the interested reader is referred to B [33], pages 241 - 248).

12. A Banach space with a sub-symmetric basis which is not a symmetric basis.

Consider the Banach space w of all real sequences $x = (x_n)$ equipped with the norm

$$||x|| = \sup \sum_{i=1}^{\infty} \left| \frac{x_{n_i}}{i} \right| ; \quad \{n_i\} \in \Theta \} < \infty$$

where

$$\Theta = \{\{n_i\} \in \mathbb{N} ; \; n_1 < n_2 \; . \; . \; . \} .$$

The unit vectors $\{\delta_i\}$, $\delta_i = (\delta_{ij})_{j=1}^{\infty}$ form a sub-symmetric basis of w which is not symmetric. (For a proof, see B [33], page 583).

13. A Banach space without a subsymmetric basis.

The Banach space $L^p[0,1]$, $1<p<\infty$, $p \neq 2$, does not have subsymmetric basis (B [33], page 563).

14. An E-complete biorthogonal system in a Banach space, which is not a basis.

Let E be the space of all continuous functions on $\mathbb{R}$ having period 2π and equipped with the norm

$$||x|| = \sup \{|x(t)| \; ; \; t \in \mathbb{R} \}.$$

Let

$$x_0(t) = 1$$

$$x_{2i-1}(t) = \sin(it)$$

$$x_{2i}(t) = \cos(it), \; t \in \mathbb{R}, \; i = 1,2, \ldots$$

and

$$f_{2i}(x) = \frac{1}{\pi} \int_{-\pi}^{\pi} x(s) \cos(is) \, ds ,$$

$$f_{2i+1}(x) = \frac{1}{\pi} \int_{-\pi}^{\pi} x(s) \sin(i+1)s \, ds, \; x \in E, \; i = 1,2 \ldots .$$

Then (x_i, f_i) is an E-complete biorthogonal system, since $\{x_i\}$ is complete in E. However, from the existence of continuous functions whose Fourier series are not uniformly convergent, it follows that $\{x_i\}$ is not a basis of E.

15. A normed space with a basis which is not a Schauder basis.

Consider the Banach space $E = C[0,1]$ as defined in # 3 and let

$$x_i(t) = t^{i-1} , \; t \in [0,1] , \; i = 1,2, \ldots .$$

Then $\{x_i\}$ is w-linearly independent (that is, $\sum_{i=1}^{\infty} \alpha_i x_i = 0$,

$\{ \alpha_i \} \subset \mathbb{R}$, implies $\alpha_i = 0$, $i = 1,2, \ldots$) but not minimal (that is, x_i does not belong to the subspace spanned by $\{x_i, \ldots, x_{i-1}, x_{i+1} \ldots\}$ by B [25(a)] , Chapter III, 3, Theorem 2). Also the span of the sequence $\{x_i\}$ is E itself.

Let

$$M = \{ \sum_{i=1}^{\infty} \alpha_i x_i \varepsilon E \; ; \; \{\alpha_i\} \subset \mathbb{K} , \; \sum_{i=1}^{\infty} \alpha_i x_i \text{ converges } \} .$$

Then M is a normed space under the induced topology from E. $\{x_i\}$ is a basis for M but not a Schauder basis for M. Infact, if there is a sequence $\{\phi_i\} \subset M'$ such that

$$\phi_i(x_j) = \delta_{ij} , \; i,j = 1,2, \ldots ,$$

then since $\overline{M} = E$, we can expand each ϕ_i, by continuity, to an $f_i \varepsilon E'$ and we have

$$f_i(x_j) = \delta_{ij} , \; i,j = 1,2, \ldots$$

which contradicts the assumption that $\{x_i\}$ is not minimal.

16. A normed space with a Schauder basis which is neither an (e)-Schauder basis nor a (b)-Schauder basis.

Consider the space E of # 14. Let (x_i, f_i) be the E-complete biorthogonal system such that $\{x_i\}$ is not a basis of E as in # 14. Consider the subspace

$$M = \{x \varepsilon E \; ; \; \lim_n T_n(x) = \lim_n \sum_{i=1}^{n} f_i(x) x_i = x \} ,$$

equipped with the induced norm topology. Then $\{x_i\}$ is a Schauder basis for M, but not a (b)-Schauder basis for M. Infact, if

$$\sup \{ ||T_n'|| \; ; \; 1 \leq n < \infty \} < \infty$$

where T'_n is T_n restricted to M, then, since $\overline{M} = E$, we have

$$\sup \{ ||T_n|| \; ; \; 1 \leq n < \infty \} < \infty .$$

Hence there is a constant $K \geq 1$ such that

$$||T_n|| \leq K, \; n = 1.2, \; . \; . \; . \; .$$

Now for every finite linear combination

$$p = \sum_{j=1}^{m} \alpha_j \, x_j$$

and all $i \geq m$, we have

$$\begin{aligned} T_n(p) &= \sum_{i=1}^{n} f_i(\sum_{j=1}^{m} \alpha_j \, x_j) \, x_i \\ &= \sum_{i=1}^{n} \sum_{j=1}^{m} \alpha_j \, \delta_{ij} \, x_i \\ &= \sum_{j=1}^{m} \alpha_j \, x_j = p \end{aligned}$$

so that $\lim_n T_n(x) = x$ for all x in E. But this contradicts the fact that $\{x_i\}$ is not a basis for E. Since an (e)-Schauder basis is equivalent to a (b)-Schauder basis for a normed space, it follows that $\{x_i\}$ is not a (b)-Schauder basis for M.

17. A Banach space whose dual has a weak* basis but no basis.

Consider the Banach space $E = \ell^1$. Then the unit vectors in $E' = \ell^\infty$ constitute a weak* basis for E' (because a sequence $\{f_i\}$ in a conjugate Banach space E' is a weak* Schauder basis for E' if and only if E has a basis $\{x_i\}$ whose coefficient functionals are $\{f_i\}$); but $E' = \ell^\infty$ has no basis since it is not separable.

18. A Banach space whose dual space has a basis which is not a weak* basis.

Consider the Banach space $E = c_0$, the space of sequences converging to 0, equipped with the supremum norm topology.

Define the sequence $\{h_i\}$ in $E' = \ell^1$ by

$$h_1 = f_1, \; h_i = \delta_{i-1} - \delta_i \;, \; i = 2,3, \dots$$

where $\delta_i = (\delta_{ij})$ is the unit vector basis of ℓ^1. The sequence $y_i = \{1, \dots, 1, 0, 0, \dots\}$, 1 occuring i times and $i = 1,2, \dots$, is a basis for E. (This follows from the following result: Let $\{x_i\}$ be a bounded basis in a Banach space E and $\{\alpha_i\}$ a sequence of scalars such that $\alpha_i \neq 0$, $i = 1,2, \dots$. The sequence $\{y_i\}$ in E defined by

$$y_i = \sum_{n=1}^{i} \alpha_n x_n \;, \; i = 1,2, \dots,$$

is a basis for E if and only if the sequence $\dfrac{||y_{i+1}||}{|\alpha_{i+1}|}$ is bounded). Identifying canonically ℓ^1 with E', the sequence $\{h_{i+1}\}$ is obviously the sequence of coefficient functionals for the basis $\{y_i\}$ of E, whence $\{h_{i+1}\}$ is a basis for the closed subspace of E' spanned by $\{h_{i+1}\}$, because if (x_i, f_i) is a basis for a Banach space E, then $\{f_i\}$ is a basis for the closed subspace $[f_i]$ of E' spanned by $\{f_i\}$ and

$$f = \sum_{n=1}^{\infty} f(x_n) f_n \;, \; f \in [f_i]).$$

Now, $h_1 \notin [h_{i+1}]$, since for the functional

$$\phi_0(f) = \sum_{n=1}^{\infty} \zeta_n , \quad f = (\zeta_i) \varepsilon E' ,$$ we have

$\phi_0(h_1) = 1$, $\phi_0(h_{i+1}) = 0$, $i = 1,2, \ldots$, and $\{h_i\}$ is complete in E' , since the relations

$$\phi \varepsilon E'', \ \phi(h_i) = 0, \ i = 1,2, \ldots$$

imply $\phi=0$. Consequently $\{h_i\}$ is a basis for E'. Furthermore, although every $f \varepsilon E'$ has a weak* expansion

$$f(x) = \sum_{n=1}^{\infty} \alpha_n h_n(x), \ x \varepsilon E,$$

this weak* expansion is not unique, since by

$$h_1(x) = \sum_{n=2}^{i} h_n(x) = f_i(x), \ x \varepsilon E, \ i = 2,3, \ldots$$

we have

$$h_1(x) - \sum_{n=2}^{\infty} h_n(x) = 0, \ x \varepsilon E.$$

Thus, $\{h_i\}$ is not a weak* basis for E' .

19. A Banach space whose dual space has a weak* basis which is not a weak* Schauder basis.

Consider the Banach space $E = c_0$ and define the sequence $\{f_i\}$ in $E' = \ell^1$ by

$$(*) \quad f_1 = \delta_1, \ f_i = (-1)^{i+1}\delta_1 + \delta_i , \ i = 2,3, \ldots$$

where $\{\delta_i\}$ is the unit vector basis of ℓ^1.

Then $\{f_i\}$ is a basis for E' in the norm topology and a weak* basis for E'. To show that it is not a weak* Schauder basis, we proceed as follows: Let $\{\phi_i\} \subset E''$ be the coefficient functionals for the basis $\{\delta_i\}$ of E'. Define

$$(**)\quad \chi_1(f) = \phi_i(f) + \sum_{m=2}^{\infty} (-1)^m \phi_m(f) ,$$

$$f \varepsilon E', \quad \chi_i = \phi_i , \; i = 2,3, \ldots .$$

Then (f_i, χ_i) is a biorthogonal system and we have, for $f \varepsilon E'$, $i = 1,2, \ldots,$

$$\sum_{n=1}^{i} \chi_n(f) f_n = |\phi_1(f) + \sum_{m=2}^{\infty} (-1)^m \phi_m(f)|\delta_1$$

$$+\sum_{m=2}^{i} \phi_m(f) \left|(-1)^{m+1} \delta_1 + \delta_m \right|$$

$$= \sum_{m=1}^{i} \phi_m(f)\delta_m + |\sum_{m=i+1}^{\infty} (-1)^m \phi_m(f)|\delta_i .$$

Hence, since $\sum_{m=1}^{\infty} |\phi_m(f)| < \infty$ $(f \varepsilon E' = \ell^1)$,

for every $\varepsilon > 0$ and $f \varepsilon E'$, there exists an integer $N > 0$ (depending on ε and f) such that

$$|| \sum_{m=1}^{i} \chi_m(f) f_m - \sum_{m=1}^{i} \phi_m(f) \delta_m|| < \varepsilon ,$$

for $i > N$. Hence

$$f = \sum_{m=1}^{\infty} \chi_m(f) f_m , \; f \varepsilon E' .$$

From this, it follows that $\{f_i\}$ is a basis for E' in the norm topology. Furthermore, it follows from the above that every $f \varepsilon E'$ also has a weak* expansion

$$f(x) = \sum_{m=1}^{\infty} \chi_m(f) f_m (x), \; x \varepsilon E .$$

We now show that this expansion is unique. Let

$$(***)\quad \sum_{m=1}^{\infty} \alpha_m f_m (x) = 0, \; x \varepsilon E,$$

for a sequence $\{\alpha_i\}$ of scalars. Since, for $x = b_i$ $(i = 1,2, \ldots)$, where $\{b_i\}$ is the unit vector basis $\{\delta_i\}$

of $E = c_0$, we have, by (*) ,

$$f_m(b_1) = (-1)^{m+1}, \; f_m(b_i) = \delta_{mi} ,$$

$$(i = 2,3, \ldots ; m = 1,2, \ldots).$$

and by (***) it follows that

$$\sum_{m=1}^{\infty} (-1)^{m+1} \alpha_m = 0, \quad \alpha_2 = \ldots = \alpha_i = \ldots = 0 ,$$

whence $\alpha_i = 0$ $(i = 1,2, \ldots)$. Thus $\{f_i\}$ is weak* basis for E'. On the otherhand, by (**) $\chi_1 \notin \pi(E)$, where π is the canonical map of E into E', and χ_i is not weak* continuous. Thus $\{f_i\}$ is not a weak* Schauder basis for E'.

20. A separable locally convex space which has no basis.

Consider the locally convex space $(\ell_\infty', \sigma(\ell_\infty', \ell_\infty))$, $\ell_\infty = \ell^\infty$. It is separable since the sequence $\{f_i\}$ in ℓ_∞' defined by

$$f_i(x) = \zeta_i , \; x = (\zeta_i) \; i = 1,2, \ldots ,$$

is total on ℓ_∞ , whence its w*-closed linear hull is ℓ_∞'. Suppose it has a weak* basis $\{f_i\}$. Then, to every f in ℓ_∞' there is a unique sequence $\{\alpha_i\}$ such that

$$f = \lim_n \sum_{i=1}^{n} \alpha_i f_i$$

in the topology $\sigma(\ell_\infty', \ell_\infty)$. But then the series converges weakly to f. Thus $\{f_i\}$ is a weak basis for ℓ_∞' and hence a basis for ℓ_∞' . But then the Banach space ℓ_∞' is separable in the strong topology and so ℓ_∞ is separable in the strong topology which is not true.

21. A basis in a locally convex space, which is not a Schauder basis.

Let E be the space of all real-valued functions expandable as absolutely summable power series on the interval $[0,1)$ with the topology of uniform convergence on compact subsets of $[0,1)$. Let $\{x_i\}$ be a sequence in E defined by

$$x_i(t) = t^i, \; i = 0,1, \ldots .$$

Then $\{x_i\}$ is a basis for E. (This follows from the following result : If two power series $\sum_{n=0}^{\infty} \alpha_n (z - z_0)^n$ and $\sum_{n=0}^{\infty} \beta_n (z - z_0)^n$ converge on a neighbourhood of z_0 absolutely to the same sum, then $\alpha_n = \beta_n$ for all n). For each polynomial P in t, the coefficient functional f_1, determined by the expansion coefficient α_1, is given by

$$f_1(P) = \lim_{t \to 0} \frac{P(t) - P(0)}{t} .$$

Since the polynomials are dense in the Banach space $C[0,1]$, we can choose a polynomial z_n in E for each n such that

$$\sup \{ |z_n(t) - (1-nt)| \; ; \; t \in [0,1/n] \} \leq \frac{1}{n}$$

and

$$\sup \{ |z_n(t)| \; ; \; t \in (\tfrac{1}{n}, 1) \} \leq \frac{1}{n} .$$

Define

$$y_n(t) = \int_0^t z_n(t')\, dt' , \; t \in [0,1) .$$

Then y_n is a polynomial. Therefore

$$f_1(y_n) = \lim_{t \to 0} \frac{y_n(t)}{t} = z_n(0) \in [\, 1 - \tfrac{1}{n} , 1 + \tfrac{1}{n} \,],$$

$(n = 1,2, \ldots)$, so that

$$\lim_n f_1(y_n) = 1 .$$

But $\lim_n y_n = 0$, since

$$\sup \{|y_n(t)| \; ; \; t \in [0,1)\} \leq \int_0^t |z_n(t)| \, dt$$

$$\leq \frac{1}{2n} + \frac{1}{n} = \frac{3}{2n} .$$

22. A complete, metrizable and separable (non-locally convex) topological vector space which has no basis.

Consider the complete, metrizable and separable (non-locally convex) topological vector space $E = L^p[0,1]$, $0<p<1$, as defined in Chapter 1, # 6. It is separable. Since $E' = \{0\}$, as shown in Chapter 1, # 6, E has no Schauder basis and hence no basis.

23. A generalized basis in a non-separable Banach space, which is not a Markushevich basis.

Let E be the Banach space of all bounded real-valued functions on an infinite set I, under the supremum norm. For each $i \in I$, define x_i as the function assuming the value 1 at i and 0 elsewhere. The point functions $f_i(x) = x(i)$ are all continuous and $\{f_i\}$ is biorthogonal to $\{x_i\}$. Since $f_i(x) = 0$ implies $x = 0$ for all x in E, it follows that $\{x_i\}$ is a (countable) generalized basis in E. It is easy to see that E is non-separable and $\{x_i\}$ is not total in E. That $\{x_i\}$ is not a Markushevich basis in now clear.

24. A Markushevich basis in a Fréchet space, which is not a Schauder basis.

Let E be the Fréchet space of all functions analytic on the open unit disc $|z| < 1$, equipped with the topology via the metric of uniform convergence on compact sets. We set

$$f_n(z) = (z-a)^n, \; n = 0,1,2 \ldots$$

where a is any complex number satisfying the inequality $0<|a|< 1$. These functions form a Markushevich basis in E, the corresponding biorthogonal sequence of continuous linear functionals being given by

$$\phi_n(f) = \frac{f^{(n)}(a)}{n!}, \quad n = 0,1,2 \ldots .$$

On the other hand, $\{f_n\}$ is not a Schauder basis, since the series $\sum_{n=0}^{\infty} \frac{(z-a)^n}{(1-a)^{n+1}}$ corresponding to the functions $f(z) = (1-z)^{-1}$ diverges for z outside the circle

$$|z-a| = |1-a| .$$

25. A maximal biorthogonal system in a Fréchet space, which is not a generalized basis.

Let E be the Fréchet space as defined in # 24. We set

$$f_0(z) = 1,$$

$$f_1(z) = 1 + z,$$

$$\ldots\ldots\ldots\ldots\ldots$$

$$f_n(z) = 1+z+ \ldots + z^n,$$

$$\ldots\ldots\ldots\ldots\ldots$$

Define ϕ_n on E by

$$\phi_n(f) = \frac{f^{(n)}(0)}{n!} - \frac{f^{(n+1)}(0)}{(n+1)!}, \; n = 0,1, \ldots .$$

Then (f_n, ϕ_n) form a biorthogonal system over E. Since $\{f_n\}$ is total in E, $\{\phi_n\}$ is the only possible sequence of coefficient functionals. But the $\phi_n(f) = 0$, $n = 0,1,\ldots$, does not imply that $f = 0$; infact the condition $\phi_n(f) = 0$,

$n = 0, 1, \ldots$ is necessary and sufficient for f to be of the form

$$f(z) = C(1 + z + z^2 + \ldots + z^n + \ldots)$$

where C is an arbitrary constant. Thus, $\{f_n\}$ is not a generalized basis in E. Now, if (f_n, ϕ_n) is not maximal, then there exists a continuous linear functional annihilating $\{f_n\}$ but not vanishing identically. But this is impossible, since every continuous linear functional ϕ on E has the representation

$$\phi(f) = \sum_{n=0}^{\infty} \frac{f^{(n)}(0)}{n!} h_n$$

for a suitably chosen sequence $\{h_n\}$ of complex numbers.

26. An extended unconditional basis in a countably barrelled space, which is not an extended unconditional Schauder basis.

Let E and E' be as in Chapter 3, # 43. Choose $\lambda_0 \varepsilon \Lambda$. Define

$$y_\lambda = (\eta_\mu) \varepsilon E, \ \mu \varepsilon \Lambda$$

by

$$\eta_\mu = \begin{cases} 1 & \text{if } \mu=\lambda \text{ or } \mu=\lambda_0. \\ 0 & \text{otherwise}. \end{cases}$$

Then $\{y_\lambda\}$ is an extended unconditional basis in E. But if

$$x = (\zeta_\lambda) \varepsilon E, \ \lambda \varepsilon \Lambda$$

then

$$x = \sum_{\lambda \neq \lambda_0} \zeta_\lambda y_\lambda + (\zeta_{\lambda_0} - \sum_{\lambda \neq \lambda_0} \zeta_\lambda) y_{\lambda_0},$$

and the coefficient functional of y_{λ_0} clearly does not belong to E'.

27. The isomorphism theorem fails if the domain or the

range space is not barrelled.

(i) Let E be a Banach space with a Schauder basis (x_n, f_n). Let F be E equipped with the weak topology. Then $\{x_n\}$ in E and $\{x_n\}$ in F are similar, but E and F are not homeomorphic.

(ii) Let (x_n, f_n) denote the unit vector basis in $E = \ell^1$. Let F be ℓ^1 considered as a dense subspace of c_0 and let (y_n, g_n) denote the unit vector basis of F. Then $\{x_n\}$ and $\{y_n\}$ are similar, but E and F are not isomorphic.

28. The isomorphism theorem does not hold for generalized basis even if the domain and the range spaces are complete and barrelled.

Let E be an infinite dimensional Banach space with a generalized basis (x_λ, f_λ), and let F be E equipped with the finest locally convex topology. Then E and F are complete barrelled spaces and (x_λ, f_λ) is a generalized basis in F. Clearly $\{x_\lambda\}$ in E and $\{x_\lambda\}$ in F are similar. But, since F is not metrizable, E and F are not homeomorphic.

29. A vector space with two compatible locally convex topologies such that there is a Schauder basis for one topology, which is not a Schauder basis for the other topology.

Let ϕ denote the space of finite sequences and e the sequence with every coordinate equal to 1. Define

$$\lambda = \{x = (x_i) \ ; \ (x_i) \text{ is eventually constant}\}$$
$$= \{\alpha e + x \ ; \ \alpha \text{ is a scalar, } x \in \phi\} \quad .$$

Clearly the Köthe dual $\lambda^{\times}$ is ℓ^1 and the sequence $(e^{(n)})$ is a $\sigma(\lambda, \lambda^{\times})$-Schauder basis for λ , where $e^{(n)}$ is the sequence with 1 in the nth place and zeros elsewhere. We now show that $(e^{(n)})$ is not a $\tau(\lambda, \lambda^{\times})$-Schauder basis for λ . Let

$$u^{(n)} = e^{(n)} - e^{(n+1)}$$

be considered as an element of $\lambda^{\times}$. Then, although for each m ,

$$\lim_{n} < e^{(n)} , u^{(m)} > = < e, u^{(m)} > ,$$

the convergence is not uniform with respect to m. Infact,

$$< e^{(n)}, u^{(n)} > - < e, u^{(n)} > = 1$$

for all n. Hence we need only to show that the convex circled hull A of the set $\{u^{(n)}\}$ is weakly relatively compact. Let S be the closed unit ball in ℓ^1 . Then

$$A = \{ a = (a_i) ; a = \sum_{j=1}^{\infty} \beta_j u^{(j)}, \beta = (\beta_j) \varepsilon S \cap Q \}.$$

Hence A is a coordinatewise bounded subset of the space w of all sequences, and since $(w, \sigma(w,\phi))$ is a Montel space, A is $\sigma(w,\phi)$-relatively compact. Furthermore, if $\alpha e + x \varepsilon \lambda$ and $a \varepsilon A$, then

$$(*) < \alpha e + x,a > = \alpha < e,a > + < x, a >$$

$$= \alpha \sum_{j=1}^{\infty} \beta_j < e, u^{(j)} > + < x, a >$$

$$= < x, a > .$$

Therefore, if $(a^{(\nu)})$ is a net in A, there is a cofinal subset which we denote by $(a^{(\nu)})$ itself, and an element

$a^{(0)} \varepsilon \omega$ such that if $\alpha e + x \varepsilon \lambda$, in view of (*) ,

$$\lim_{\nu} < \alpha e + x, a^{(\nu)} > = \lim_{\nu} < x, a^{(\nu)} >$$

$$= < x, a^{(0)} > \ .$$

Now all that we need to show is that $a^{(0)} \varepsilon \lambda^{\times} = \ell^1$ and that (*) remains valid if a is replaced by $a^{(0)}$, for this means that $a^{(0)}$ is a $\sigma(\lambda^{\times}, \lambda)$ -limit point of the original net. Since $A \subset 2S$, $a^{(\nu)} \varepsilon 2S$ for all ν. Since $a^{(0)}$ is the coordinatewise limit of the net ($a^{(\nu)}$),

$$a^{(0)} \ \varepsilon \ 2S \subset \ell^1 = \lambda^{\times} \ .$$

Furthermore, we may write

$$a^{(\nu)} = \sum_{j=1}^{\infty} \beta_j^{(\nu)} u^{(j)} \ , \quad \beta^{(\nu)} = (\beta_j^{(\nu)}) \varepsilon \ S \cap \phi \ .$$

Hence

$$a_i^{(\nu)} = \beta_i^{(\nu)} - \beta_{i-1}^{(\nu)} \ , \quad (\beta_0^{(\nu)} = 0)$$

and so

$$\beta_i^{(\nu)} = a_1^{(\nu)} + \ . \ . \ . + a_i^{(\nu)}$$

and since ($a^{(\nu)}$) is coordinatewise convergent to $a^{(0)}$, we may define

$$\beta_i^{(0)} = \lim_{\nu} \beta_i^{(\nu)} = a_1^{(0)} + \ . \ . \ . + a^{(0)} \ .$$

Therefore, $(\beta^{(\nu)})$ is a net in S, which is coordinatewise convergent to $\beta^{(0)} = (\beta_i^{(0)})$

and so

$$\beta^{(0)} \ \varepsilon \ S \subset \ell^1 \ .$$

Hence

$$\lim_i \beta_i^{(0)} = 0 .$$

Therefore,

$$\begin{aligned} < \alpha e + x, a^{(0)} > &= \alpha < e, a^{(0)} > + < x, a^{(0)} > \\ &= \alpha \sum_{i=1}^{\infty} a_i^{(0)} + < x, a^{(0)} > \\ &= \alpha \lim_i \beta_i^{(0)} + < x, a^{(0)} > \\ &= < x, a^{(0)} > . \end{aligned}$$

Hence $a^{(0)}$ satisfies (*) .

CHAPTER 8

TOPOLOGICAL ALGEBRAS

Introduction

A subset S of an algebra is called a left (right) ideal of A if S is a vector subspace of A such that $xy \in S$ (respectively $yx \in S$) for all $x \in A$ and $y \in S$. S is called a two-sided ideal if it is both left and right ideal. S is called maximal if it is different from A and is not properly contained in any ideal of the same type except A. A left ideal L in an algebra A is regular (or modular) if there exists $e \in A$ such that $A(1-e) \subset L$; in other words, the element e is a right identity for A modulo L. Similarly a right ideal R is regular (or modular) if there exists a left identity for A modulo R. A two-sided ideal I is regular if it is regular both as a left and as a right ideal. The intersection of all the maximal ideals in A is called a radical. If the radical R is $\{0\}$, then A is called semi-simple. A subset S in an algebra A is called idempotent if $SS \subseteq S$ and m-convex if it is idempotent and convex. The circle operation in A is defined by $x \circ y = x+y-xy$. An element of A which has a left (right) inverse relative to the circle operation is said to be left (right) quasi-regular. If r is both left and right quasi-regular, then it is called quasi-regular.

The set of elements of an algebra whose inverses exist in A, is denoted by G(A).

A normed space over $\mathbb{K}$ which is also an algebra is called a normed algebra if $||xy|| \leq ||x|| \; ||y||$. If the underlying space is a Banach space, then it is called a Banach algebra. It is commutative if $xy = yx$ for all $x,y \in A$. An element e of a Banach algebra is called identity if $||e|| = 1$. A family $\{e_\alpha ; \alpha \in I\}$ of elements of a Banach algebra A where I is a directed set, is called an approximate identity if $||e_\alpha|| \leq 1$ for each α and $\lim_\alpha e_\alpha x = \lim_\alpha x e_\alpha = x$ for each $x \in A$.

A Banach algebra A is called a Banach *-algebra if it has an involution, that is, if there exists a mapping $x \to x^*$ of A into itself with the following properties:

(a) $(x+y)^* = x^* + y^*$;

(b) $(\lambda x)^* = \bar{\lambda} x^*$;

(c) $(xy)^* = y^* x^*$;

(d) $x^{**} = x$.

A B*-algebra is a Banach *-algebra which satisfies $||x^* x|| = ||x||^2$. An A* -algebra is a Banach *-algebra in which there is defined a second norm $|||x|||$ which satisfies, in addition to the multiplicative condition $|||xy||| \leq |||x||| \; |||y|||$, the B*-condition $|||x^*x||| = |||x|||^2$. The second norm is called an auxiliary norm. A Banach *-algebra A is called symmetric if every element of the form $-x^* x$ is quasi-regular in A.

A locally convex space A which is also an algebra is called a locally convex algebra if the ring multiplication in A is separately continuous. A complete metrizable locally convex algebra is called a Fréchet algebra. A locally convex algebra A is called locally m-convex if there is a basis of neighbourhoods at 0 consisting of m-convex and circled sets. The set of all non-zero multiplicative linear functionals of A is denoted by $m^{\#}(A)$ and continuous members of $m^{\#}(A)$ by $m(A)$. If $m(A)$ is non-empty, we topologize it as follows: Let A' have the weak topology and let $m(A)$ have the relative topology it inherits as a subset of A'. With this topology, $m(A)$ is a completely regular Hausdorff space. We define a map $G: A \to C(m(A))$ by $G(x)(f) = f(x)$ for each $x \in A$ and $f \in m(A)$. G is called the Gelfand map.

An element x of a locally m-convex algebra A is called a left topological divisor of zero if $\inf \{p_\alpha(xy) ; p_\alpha(y) = 1\} = 0$. Similarly we can define right and two-sided topological divisors of 0.

Let $\mathcal{M}$ be a class of locally m-convex algebras and $A \in \mathcal{M}$. Let $x \in A$. Then x is said to be $\mathcal{M}$-singular if it is singular (that is, the inverse of x does not exist) in any superalgebra $B \supset A$ belonging to the class $\mathcal{M}$.

A locally m-convex algebra A is called a P-algebra if $\{x ; x^n \to 0\}$ is a neighbourhood of 0. A locally convex algebra A is a Q-algebra if the set of quasi-regular elements of A is open in A. A is a Q-algebra if and only if the set of quasi-regular elements of A has an interior.

A subset S of a locally m-convex algebra A is said to be m-bounded if for some $\lambda > 0$, λS is contained in a bounded and idempotent set. A locally m-convex algebra A is said to be p.i.b. if for all $x \in A$, $\{x\}$ is m-bounded. Any normed algebra is p.i.b.. A locally m-convex algebra is called m-barrelled if every m-barrel (that is, idempotent barrel) is a neighbourhood of 0. A locally m-convex algebra is called countably m-barrelled if every m-barrel which is the countable intersection of circled, convex and closed neighbourhoods of 0 is itself a neighbourhood of 0.

Let A be a locally convex algebra. An element $x \in A$ is said to be bounded if for some non-zero complex number λ, the set $\{(\lambda x)^n ; n = 1,2, \ldots\}$ is a bounded subset of A. The set of all bounded elements of A is denoted by A_0 . We write $\mathfrak{B}_1$ to denote the family of all subsets S of A such that S is circled, convex, bounded, closed, $e \in S$ and $S^2 \subseteq S$. For each S in $\mathfrak{B}_1$, let A(S) denote the subalgebra of A generated by S. Then,

$$A(S) = \{\lambda x ; \lambda \in C , x \in S\}$$

and

$$||x||_S = \inf \{\lambda > 0 ; x \in \lambda S\} , x \in A(S)$$

defines a norm on A(S) which makes A(S) a normed algebra. A locally convex algebra A is pseudo-complete if each of the normed algebras A(S), $S \in \mathfrak{B}_1$, is a Banach algebra. If A is sequentially complete, then it is pseudo-complete.

PROPOSITION 1. If $\mathfrak{B}_1$ contains a basic sub-division $\mathfrak{B}_2$ (i.e. for every $B_1 \in \mathfrak{B}_1$ there is some $B_2 \in \mathfrak{B}_2$ such that $B_1 \subseteq B_2$) such that A(S) is a Banach algebra for every $S \in \mathfrak{B}_2$, then A is pseudo-complete. (Cf. P [1])

A locally convex algebra A with a continuous involution $x \to x^*$ is called a locally convex *-algebra. With identity e, it is said to be symmetric if, for every $x \in A$, $e + x^*x$ has a bounded inverse (that is, it has an inverse belonging to A).

A pseudo-complete locally convex $*$-algebra A with identity e is called a GB*-algebra if A is symmetric and $\mathcal{B}^* = \{S \in \mathcal{B} \ ; \ S = S^*\}$ has a greatest number.

Let A be a locally convex space which is also an algebra. A subset V of A is said to be left (multiplicatively) absorbing if aV is absorbed by V for every a in A. It is right (multiplicatively) absorving if Va is absorbed by V for every a in A. It is (multiplicatively) absorbing (m-absorbing) if it is both left and right absorbing. Let p and q be two members of the family P of semi-norms generating the topology of A. p is said to absorb q if there exists a positive real number M such that $q(x) \leq Mp(x)$ for every x in A. The semi-norms p and q are said to be conjugate if they are mutually absorbing. The left-translate ${}_ap$ (right-translate p_a) of a semi-norm $p \in P$ by the element a of A is defined by ${}_ap(x) = p(ax)$ {respectively, $p_a(x) = p(xa)$} for x in A. A semi-norm p is left (right) absorbing if it absorbs all of its left (right) translates, and absorbing if it is both left and right absorbing. A is called an A-convex (absorbing convex) algebra if there exists a family P of absorbing semi-norms defining the topology of A. Every locally m-convex algebra is an A-convex algebra.

A p-normed space A which is also an algebra is called a p-normed algebra (locally bounded algebra) if the p-norm $||\cdot||$, $0<p\leq 1$. satisfies the following:

$$||xy|| \leq ||x|| \ ||y|| \ .$$

A locally semi-convex space which is also an algebra is called a locally m-semi-convex algebra if its topology is generated by a family $\{q_\alpha\}$ of k_α-seminorms satisfying the following:

$$q_\alpha(xy) \leq q_\alpha(x)\, q_\alpha(y).$$

1. An algebra which can not be made into a Banach algebra.

Consider the algebra $A = \mathcal{D}[0,1]$ of infinitely differentiable complex functions on $[0,1]$. There is no norm on A which makes it a Banach algebra. For, otherwise there would exist, for any integer $m>0$, a number c_m such that

$$\sup \{|f^{(m)}(x)| \; ; \; x\varepsilon [0,1]\} \leq c_m ||f||$$

for all $f\varepsilon A$. But we can construct a function f satisfying

$$|f^{(m)}(0)| \geq mc_m$$

for all m.

2. A Banach algebra which has no radical.

(i) The Banach algebra $\mathcal{D}^m[0,1]$ of complex functions m times continuously differentiable on $[0,1]$ with the usual pointwise multiplication and norm defined by

$$||f|| = \sum_{p=0}^{m} \frac{1}{p!} \sup |f^{(p)}(x)| ,$$

has no radical.

(ii) The Banach algebra C(X) of continuous functions on a compact space X with the usual supremum norm and pointwise multiplication has no radical.

3. A Banach algebra with a closed ideal which is not an

intersection of maximal regular ideals.

Let G be a non-compact locally compact abelian group. Then the Banach algebra $L^1(G)$ with convolution as multiplication contains a closed ideal which is not an intersection of maximal regular ideals. (See P [62]) .

4. A Banach algebra with an approximate indentity which is not an identity.

(i) Let G be a non-discrete topological group. The Banach algebra $L^1(G)$ with convolution as the multiplication does not have an identity, because $L^1(G)$ will have an identity if and only if G is discrete. But $L^1(G)$ does always contain an approximate identity, namely the family $\{e_V ; V \in \upsilon\}$ where υ is the set of all compact neighbourhoods of the identity in G and e_V is any non-negative real function on G which vanishes outside V and for which

$$\int e_V(t)\, dt = 1.$$

The partial ordering $V_1 \leq V_2$ in υ is defined by the inclusion $V_2 \subset V_1$. We then have

$$\lim_V (e_V * f) = \lim_V (f * e_V) = f$$

for every f in $L^1(G)$.

(ii) Any B*-algebra is also an example of a Banach space with approximate identity consisting of hermitian elements, which is not an identity. (See P [83]).

5. An A*-algebra which is not a B*-algebra.

Consider the Banach algebra $L^1(G)$, where G is a locally compact topological group and convolution is the multiplication. With an involution $x \to x^*$ defined by

$$x^*(t) = \Delta(t^{-1})\,\overline{x(t^{-1})},\ t\varepsilon G,$$

$L^1(G)$ is an A*-algebra which is not a B*-algebra.

6. An A*-algebra which is not symmetric.

Let G be the group of all complex 2×2 matrices (α_{ij}) with determinant equal to 1. Let υ be the subgroup consisting of unitary matrices in G and denote by M the subalgebra of $L^1(G)$ consisting of all f in $L^1(G)$ such that

$$f_\delta = f^\delta = f,\ \delta\varepsilon\upsilon\ ,\ \text{where } f_\delta(x) = f(\delta^{-1}x),$$

$f^\delta(x) = f(x\delta^{-1}), x\varepsilon G.$

Then M is a closed and commutative *-subalgebra of $L^1(G)$. Furthermore, M contains quasi-inverses, that is, if $f\varepsilon M$ and f is quasi-regular in $L^1(G)$, then it is quasi-regular in M. But M is not symmetric so that $L^1(G)$ is not symmetric.

7. A Fréchet algebra which is not a Banach algebra.

Consider the algebra $\mathcal{D}[0,1]$ as defined in # 1, with seminorms

$$p_n(t) = \sup\ \{|f^{(n)}(x)|\ ;\ x\varepsilon\ [0,1]\}\ .$$

We replace this system by an equivalent system of seminorms given by

$$q_n(f) = 2^n \sup_{0\le i\le n} \{2^i\} \sup_{0\le x\le 1} \{\ |f^{(i)}(x)|\ \}.$$

We then have

$$\sup_x |f^{(i)}(x)| \le \frac{q_n(f)}{2^{n+1}}$$

and so it is easy to see that

$q_n(fg) \leq q_n(f)\, q_n(g)$.

Hence $\mathcal{D}[0,1]$ is a locally m-convex and Fréchet algebra. But it is not a Banach algebra as shown in # 1.

REMARK. This is also an example of a Q-algebra which is not a Banach algebra.

8. A Fréchet algebra which is not a locally m-convex algebra.

Let L^ω denote the set of equivalence classes of measurable functions on [0,1] such that

$$p_n(f) = \left(\int_0^1 |f(x)|^n \, dx\right)^{1/n} < \infty , \quad n = 1,2, \ldots .$$

Then L^ω is a Fréchet algebra with pointwise multiplication and the topology generated by the semi-norms $\{p_n\}$. We show that L^ω is not a locally m- convex algebra. For this, it is sufficient to prove that the inverse is not continuous in L^ω, because a locally m-convex algebra is an algebra with a continuous inverse. We can find a sequence of real numbers $a_n > 0$, $a_n \to \infty$ such that

$$f_n(x) = a_n \, \Psi_A(x) + 1,$$

where

$$A = \left[\frac{1}{2} - \frac{1}{n} , \ \frac{1}{2} + \frac{1}{n} \right]$$

and

$$\Psi_A(x) = \begin{cases} 1 & \text{if } x \in A \\ 0 & \text{if } x \notin A, \end{cases}$$

is divergent in L^ω. On the otherhand $f_n \in G(L^\omega)$ and $f_n^{-1} \to e$.

9. A locally m-convex algebra which is not metrizable.

Let $\mathcal{D}$ denote the space of all infinitely differentiable functions with compact support in $\mathbb{R}^n$. If G is an arbitrary compact subset of $\mathbb{R}^n$, we denote by $\mathcal{D}_G$ the subspace of $\mathcal{D}$ consisting of those functions in $\mathcal{D}$ with compact support in G. If $k = \{k_1, \ldots, k_n\}$ is an arbitrary set of n non-negative integers, define the differential operator D^k by

$$D^k f = \frac{\partial^{|k|} f}{\partial^{k_1} x_1 \, \partial^{k_2} x_2 \ldots \partial^{k_n} x_n}$$

where

$$|k| = k_1 + \ldots + k_n.$$

The family of semi-norms $\{p_G^{(m)} ; m = 1, 2, \ldots \}$, defined by

$$p_G^{(m)}(f) = \sup \{|D^k f(x)| ; x \varepsilon G, 0 \leq |k| \leq m \},$$

generates a topology u_G under which $\mathcal{D}_G$ is a Fréchet space. The so called Schwartz topology u on $\mathcal{D}$ is the inductive limit topology on $\mathcal{D}$ with respect to the family

$\{\mathcal{D}_{G_r} ; r = 1,2, \ldots \}$ of subspaces of $\mathcal{D}$ where

$$G_r = \{x = (x_s) \varepsilon \mathbb{R}^n ; |x_s| \leq r, s = 1,2, \ldots, n \}.$$

$(\mathcal{D}, u)$ is a locally m-convex algebra with the pointwise multiplication, but is not metrizable.

10. A Fréchet algebra which has closed ideals but not closed maximal ideals.

Consider the Fréchet algebra $A = L^\omega$ as defined in # 8. It has closed ideals; for example

$$I = \{f \in L^{\omega} ; f(t) = 0, t \geq \frac{1}{2}\}$$

is a closed ideal in L^{ω}. But L^{ω} does not have closed maximal ideals in view of the Gelfand-Mazur theorem for Fréchet algebras, since if M is a maximal ideal, then A/M is a field, and if I is a closed ideal, then A/I is a Fréchet algebra.

11. A Fréchet algebra which does not have the Wiener property.

A locally convex algebra A is said to have the Wiener property if for every $f \in m(A)$, $f(x) \neq 0$, we have $x \in G(A)$. Let

$$\ell^{1+} = \bigcap_{p>1} \ell^{p}$$

where

$$\ell^{p} = \{ x = (x_n) ; \sum_{n=0}^{\infty} | x_n|^{p} < \infty\}.$$

It is a Fréchet algebra with convolution as multiplication and with the semi-norms

$$Q_p(x) = \left(\sum_{n=0}^{\infty} |x_n|^{p}\right)^{1/p} .$$

But ℓ^{1+} does not have the Wiener property. (Cf. B [41]).

12. A semi-simple locally m-convex Fréchet algebra which is a projective limit of Banach algebras which are not semi-simple.

Consider the commutative and semi-simple Banach algebra $A_0 = \ell^1$ with identify e, norm

$$||x|| = \sum_{0}^{\infty} |x_n|$$

and convolution as multiplication. Define

$$||x||_0 = \sum_0^\infty \frac{1}{2^n} |x_n| .$$

Then $||x||_0$ is a continuous submultiplicative norm on A_0 such that the completion of A_0 in the norm $||x||_0$ is not a semi-simple algebra. Now we define A as an algebra of sequences $x = (\xi_n)$, $\xi_n \varepsilon A_0$, such that

$$p_i(x) = \sup \{||\xi_1|| , \dots , ||\xi_i|| , ||\xi_{i+1}||_0 , ||\xi_{i+2}||_0 \dots \} < \infty ,$$

$$(i = 1,2, \dots).$$

Then A is a semi-simple locally m-convex Fréchet algebra with the semi-norms $\{p_i\}$ and pointwise multiplication. Let $\{q_i\}$ be an increasing system of continuous submultiplicative semi-norms on A equivalent to $\{p_i\}$. There are constants C_1 and C_2 and integers i and m such that

$$p_1(x) \leq C_1 \; q_i(x) \leq C_2 \; p_m(x)$$

for all x in A. Since $p_1(x)$ is a norm on A, $q_i(x)$ is also a norm on A. Let A_i be the completion of A in this norm. Let x_0 be a non-zero element of A_0 such that

$$\lim ||x_0^n||_0^{1/n} = 0.$$

Let

$$x = (0, \dots , x_0, x_0 , \dots) \varepsilon A ,$$

where the zeros appear in i places. We have

$$q_i(x^n) \leq \frac{C_2}{C_1} p_m(x^n) = \frac{C_2}{C_1} ||x_0^n||_0 .$$

So,

$$\lim |q_i(x^n)|^{1/n} = \lim |\frac{C_2}{C_1} ||x_0^n||_0|^{1/n} = 0$$

and $x \neq 0$. Hence A_i is not a semi-simple algebra.

13. $\mathcal{M}$-singular elements of a locally m-convex Fréchet algebra, which are not topological divisors of zero.

Consider the locally m-convex Fréchet algebra $\mathfrak{X}$ of power series

$$x = \sum_0^\infty x_n t^n$$

with semi-norms

$$p_n(x) = \sum_{i=1}^n |x_i|$$

and with Cauchy multiplication of the series, that is,

$$xy = z = \Sigma z_n t^n$$

where

$$z_k = \sum_{i=0}^k x_i y_{k-i} .$$

Writing

$$x(t) = \Sigma x_n t^n ,$$

$x_0(t) = t$ is a generator, and

$$m(A) = m^{\#}(A)$$

since A is a Q-algebra. But $x_0 - \lambda e$ is invertible in $\mathfrak{X}$ for any $\lambda \neq 0$ and $\sigma(x\ ) = \{0\}$. If $f \varepsilon m(A)$, then $f(x_0) = 0$ and so there is only one element in $m^{\#}(A)$ given by $f(x)=x(0)$ for any $x \varepsilon \mathfrak{X}$. Thus there is only one maximal ideal in $\mathfrak{X}$. Let us denote it by M. Since

$$G(\mathfrak{X}) = \mathfrak{X} - M,$$

the only topological divisors of zero can be in M. But $M = x_0\mathfrak{X}$ and so if there are topological divisors of zero in M, then x_0 is also one such divisor. But x_0 is not a topological divisor of zero in $\mathfrak{K}$, since for any $y \varepsilon \mathfrak{K}$,

$$p_n(yx_0) = p_{n-1}(y) \quad , n \geq 1.$$

Thus, $\mathfrak{K}$ has no topological divisors of zero. On the other hand,

$$M = rad(A) \subset rad(B)$$

for any m-convex superalgebra B of A, and so any element $x \varepsilon M$ must be $\mathcal{M}$-singular.

14. A locally m-convex Fréchet algebra which has neither topological divisors of zero nor $\mathcal{M}$-singular elements.

The locally m-convex Fréchet algebra ε of all entire functions of one complex variable with pointwise multiplication and semi-norms

$$p_n(x) = \sup \{x(t) \; ; \; |t| \leq n \}$$

fulfills the requirement.

15. An m-barrelled algebra which is not barrelled.

Consider the algebra A of all polynomials X without constant term, equipped with the strongest locally m-convex topology. Let α be a rational number with $0<\alpha<1$, and let $V(\alpha)$ be the circled and convex envelope of $\{\alpha^n X^n \; ; \; n$ is positive integer$\}$. Then the family of all such $V(\alpha)$ is a neighbourhood basis at 0 for the strongest locally m-convex topology on A. It is clear that A is an m-barrelled algebra. To show that it is not barrelled, let S be the circled and convex envelope of $\{2^{-n^2} X^n \; ;$ $n = 1,2, \ldots\}$.

S is absorbing and so $\overline{S}$ is a barrel. If $\beta>0$,

$$\beta 2^{-n^2} X^n \notin \overline{S} \quad ;$$

for, if

$$0<\alpha<2^{-n}(\beta-1)^{1/n}$$

Then

$$|\beta 2^{-n^2} X^n + V(\alpha)| \cap S = \emptyset .$$

For any $\alpha > 0$,

$$\lim 2^{-n^2} \alpha^{-n} = 0, \; n\to\infty .$$

So, for large n,

$$\alpha^n > 2^{-n^2} ,$$

that is,

$$\alpha^n = \beta_{\alpha,n} \; 2^{-n^2} , \; \beta_{\alpha,n} > 1 .$$

Hence, if $V(\alpha) \subset \bar{S}$, for suitably large n we have

$$\alpha^n X^n = \beta_{\alpha,n} \; 2^{-n^2} X^n \varepsilon V(\alpha) \subseteq \bar{S}$$

which is a contradiction. Hence $\bar{S}$ is not a neighbourhood of O.

16. A countably m-barrelled algebra which is not m-barrelled.

The locally convex space C(W), as defined in Chapter 3, # 41 (ii) is infact a locally m-convex algebra with the usual pointwise multiplication. It is a countably m-barrelled algebra becuase it is countably barrelled. But it is not m-barrelled because the set

$$B = \{f\varepsilon C(W) \; ; \; |f(x)| \leq 1, \; x\varepsilon W\}$$

is closed, circled, m-convex and absorbing but not a neighbourhood of O.

17. A complete p.i.b. algebra which is not a P-algebra.

The locally convex algebra C(W), as defined in

Chapter 3, # 41(ii) is complete because W is locally compact, p.i.b. because W is pseudo-compact, but not a P-algebra because W is not compact.

18. A metrizable p.i.b. algebra which is neighter a P-algebra nor an m-bornological algebra.

Let X be a non-compact completely regular space. Let E be the algebra of all continuous real-valued functions on X with compact supports, equipped with the compact-open topology u. Let v be the topology defined by the uniform norm. Then

$$B = \{f\varepsilon E \ ; \ |f(x)| \leq 1 \text{ for all } x\varepsilon X \}$$

is bounded and idempotent for u and v, and also every bounded and idempotent set for either topology is contained in B. Hence the m-bounded sets for u and v coincide. By the hypothesis on X, v is strictly stronger than u so that (E,u) is not m-bornological, because (E,u) is m-bornological if and only if no strictly stronger locally m-convex topology on E can have the same m-bounded sets. Clearly (E,u) is p.i.b.. But it is not a P-algebra, for, given any compact set K in X, there exists $f\varepsilon E$ such that $f = 0$ on K, but $f(x) > 1$ for some $x\varepsilon X$; hence the sequence $\{x^n\}$ does not converge to 0. Now let

$$X = \bigcup_{1}^{\infty} K_n ,$$

where $\{K_n\}$ is a fundamental sequence of compact subsets of X. Then E is metrizable.

19. The Gelfand map which is continuous for a locally convex algebra which is not m-barrelled.

The algebra $C(\mathbb{R})$ of complex-valued continuous functions on $\mathbb{R}$, equipped with compact-open topology, is a locally m-convex algebra. The subspace $C_b(\mathbb{R})$ consisting of bounded functions in $C(\mathbb{R})$ is not m-barrelled, since the set

$$B = \{f \in C_b(\mathbb{R}) \ ; \ |f(x)| \leq 1, \ x \in \mathbb{R}\}$$

is an m-barrel but not a neighbourhood of 0. But the Gelfand map G is continuous.

20. A GB^*-algebra which is not a locally m-convex algebra.

We have seen in # 8 that $A = L^\omega$ is not a locally m-convex algebra. We now show that it is a GB^*-algebra. Clearly it is a complete (and hence pseudo-complete) locally convex *-algebra, the involution being complex conjugation. A_0 is simply the subalgebra L^∞ of essentially bounded measurable functions, and $\mathcal{B}^*$ has the greatest member

$$B_0 = \{f \in A \ ; \ \text{ess sup } |f(x)| \leq 1\} .$$

For any f in A,

$$(e + f^* f)^{-1} = (e + |f|^2)^{-1} \in A_0$$

so that A is symmetric. Thus $A = L^\omega$ is a GB*-algebra.

21. A GB*-algebra on which there are no non-trivial multiplicative linear functionals.

Consider the GB*-algebra $A = L^\omega$ as defined in # 8. Let ϕ be a non-zero multiplicative linear functional on A. Then ϕ is also multiplicative restricted to the subalgebra $C[0,1]$. Thus, there is some point $x_0 \in [0,1]$ such that $\phi(f) = f(x_0)$ for every f in $C[0,1]$ – ϕ can not annihilate

$C[0,1]$, since $C[0,1]$ contains the identity of A. But there is certainly some continuous extended complex-valued function $h \varepsilon A$ such that

$$h(x_0) = \infty, \ h(x) \geq 0 \text{ for all } x.$$

But then there is some f_n in $C[0,1]$, $n = 1,2, \ldots$, such that $f_n(x) = 1$ and

$$n f_n(x) \leq h(x) \text{ for all } x.$$

Thus

$$\phi(h) \geq n, \ n = 1,2, \ldots$$

which is a contradiction.

22. A pseudo-complete locally convex algebra which is not sequentially complete.

Let A be the algebra of all complex polynomials P and let A be given the topology u of uniform convergence on the compact subsets of $\mathbb{R}^+$. Then A_0 consists of the constant functions and $\mathcal{B}_1$ has a greatest member, namely the set of all constant functions not exceeding unity in absolute value. Thus A is pseudo-complete, by Proposition 1. Now, let the sequence $\{P_n\}$ in A be defined by

$$P_n(x) = \sum_{r=0}^{n} \frac{(-1)^{r+1} x^{2r+1}}{(2r+1)!} .$$

Then $P_n(x)$ converges to $\sin(x)$ uniformly on compact subsets of $\mathbb{R}^+$. Let

$$B = \{P \varepsilon A \ ; \ |P(x)| \leq e^x \ (x \geq 0)\} .$$

Then B is u-bounded. Also

$$e^{-x}|P_n(x) - \sin(x)| \to 0$$

uniformly on $\mathbb{R}^+$; it is thus clear that $\{P_n\}$ is Cauchy in the sense of Mackey but does not converge to an element of

A. Thus A is not Mackey-complete which implies that A is not sequentially complete.

23. An A-convex algebra which is not a locally m-convex algebra.

(i) Let $C_b(\mathbb{R})$ be as in ≠ 19. We denote by $C_0^+(\mathbb{R})$ the set of strictly positive continuous real-valued functions on $\mathbb{R}$ which vanish at infinity. The family of semi-norms $\{p_\phi \ ; \ \phi \in C_0^+(\mathbb{R})\}$ generates a locally convex topology u on $C_b(\mathbb{R})$ where

$$p_\phi(f) = \sup \{|f(x)\ \phi(x)| \ ; \ x \in \mathbb{R}\} \ , \ f \in C_b(\mathbb{R}) \ .$$

The space $(C_b(\mathbb{R}), u)$ is A-convex and each p_ϕ fails to be submultiplicative. Suppose $(C_b(\mathbb{R}), u)$ is locally m-convex and let Q be a set of submultiplicative semi-norms which define u. We may assume that

$$\max (q_1 \ , \ . \ . \ . \ q_n) \in Q \text{ and } \lambda q_1 \in Q$$

whenever

$$q_1 , \ . \ . \ . \ q_n \in Q \text{ and } \lambda \geq 1.$$

Thus, for $\phi \in C_0^+(\mathbb{R})$, there exists $q \in Q$ and $\Psi \in C_0^+(\mathbb{R})$ such that

(*) $V(\Psi) \subseteq V(q) \subseteq V(\phi)$

where

$$V(\Psi) = \{f \ ; \ p_\psi(f) \leq 1\} \ .$$

Since $V(\Psi) \subseteq V(\phi)$, $\phi \leq \Psi$ (pointwise).

Let $\theta \in \mathbb{R}$ with

$$0 < \theta < \min (1, M(\Psi)),$$

where $M(\Psi)$ is the maximum of $|\Psi|$. Then, for some $x \in \mathbb{R}$, it follows that

$$\Psi(x) = \theta \geq \phi(x) \text{ and } \theta^{n} < \phi(x)$$

where n is a positive integer. Consider the function f(y) defined by

$$f(y) = \begin{cases} \dfrac{y - x + 1}{\Psi(y)} & \text{if } x-1 \leq y \leq x \\ \dfrac{- y + x + 1}{\Psi(y)} & \text{if } x<y \leq x+1 \\ 0 & \text{otherwise.} \end{cases}$$

Then f is well-defined, since $\Psi \varepsilon C_0^+(\mathbb{R})$ and $f \varepsilon C_b(\mathbb{R})$. But $P_\Psi(f) = 1$ and

$$p_\phi(f^n) \geq f^n(x)\ \phi(x) = \theta^{-n}\ \phi(x) > 1$$

which contradicts (*) since q is sub-multiplicative. Hence $(C_b(\mathbb{R}), u)$ is not locally m-convex.

(ii) Consider the algebra $C[0,1]$ of all continuous complex-valued functions on the closed interval $[0,1]$. A norm $||\cdot||$ is defined on this algebra by

$$||f|| = \sup \{|f(x)\ \phi(x)|\ ;\ x \varepsilon [0,1]\}$$

where

$$\phi(x) = \begin{cases} x & \text{if } 0 \leq x \leq \frac{1}{2} \\ 1-x & \text{if } \frac{1}{2} \leq x \leq 1 . \end{cases}$$

Then ($C[0,1]$, $||\cdot||$) is a normed space which is A-convex but not locally m-convex.

24. A p-normed (locally bounded) algebra which is not a normed algebra.

The algebra ℓ^p, $0<p<1$, of all complex two-sided sequences $x = (x_n)_{-\infty}^{\infty}$ with the p-norm

$$||x|| = \sum_{n=-\infty}^{\infty} |x_n|^p < \infty$$

and convolution as multiplication is a p-normed algebra but not a normed algebra.

25. A locally m-semi-convex algebra which is not a locally m-convex algebra.

Consider the algebra ℓ^{q+} , $0\leq q<1$, of all complex two-sided sequences $x = (x_n)_{-\infty}^{\infty}$ with convolution as multiplication.

Define

$$||x||_p = \sum_{n=-\infty}^{\infty} |x_n|^p$$

for every p satisfying $q<p<1$. Then ℓ^{q+} is a locally m-semi-convex algebra for the family of p-semi-norms $\{||\cdot||_p\}$, $q<p<1$, which is neither locally m-convex nor locally bounded.

OPEN PROBLEMS:

1. (Michael B [23]): If A is a complete locally m-convex algebra, is $\mathrm{R}(A)= \cap\{$closed, regular, maximal, right ideals in A$\}$ closed?

2. (Michael B [23]) : Is every multiplicative linear functional on a commutative Fréchet algebra continuous?

BIBLIOGRAPHY

BOOKS

[1] BANACH, S. : Théorie des opérations linearies, Warsaw,(1932).

[2] BIRKHOFF, G. : Lattice theory, Amer. Math. Soc. Colloq. Publ. Vol. 25 (Third Printing), (1973).

[3] BOURBAKI, N. : Éléments de mathématique, Livre V, Espaces Vectoriels topologiques, Hermann, Paris, (1953, 1955).

[4] DAY, M.M. : Normed linear spaces, Springer-Verlag, Berlin, (1958).

[5] DUNFORD, N. and SCHWARTZ, J. : Linear operators, Vol. I, Interscience, New York, (1958).

[6] EDWARDS, R.E. : Functional analysis, Theory and applications, Holt, Rinehart and Winston, New York, (1965).

[7] GELBAUM, B.R. and OLMSTED, J.M.H. : Counterexamples in analysis, Holden-Day, Inc. San Francisco, (1964).

[8] GILLMAN, L. and JERISON, M. : Rings of continuous functions, Van Nostran, Princeton, (1960).

[9] GROTHENDIECK, A.: Products tensoriels topologiques et spaces nucleaires, Mem. Amer. Math. Soc. 16 (1955).

[9(a)] GROTHENDIECK, A.: Topological vector spaces (English edition), Gordon and Breach, Sc. Publishers, New York - London - Paris, (1973).

[10] GUICHARDET, A.: Special topics in topological algebras, Gordon and Breach, Sc. Publishers, New York-London-Paris, (1968).

[11] HALMOS, P.R.: Introduction to Hilbert space and the Theory of Spectral Multiplicity, Chelsea, New York, (1951).

[12] HARDY, G.H.; LITTLEWOOD, J.E. and POLYA, G: Inequalities, Camb. Univ. Press, (1934).

[13] HORVÁTH,J.: Topological vector spaces and distributions, Vol. I, Addison-Wesley, Reading, Mass., (1966).

[14] HUSAIN, T." The open mapping and closed graph theorems in topological vector spaces, Oxford Mathematical Monographs, Clarendon Press, Oxford, (1965).

[15] HUSAIN, T. and KHALEELULLA , S.M.: Barrelledness in topological and ordered vector spaces, Lecture Notes in Mathematics, Springer-Verlag, Berlin-Heidelberg - New York, (1978).

[16] JAMESON, G.: Ordered linear spaces, Lecture Notes in Mathematics, No. 141, Springer-Verlag, Berlin-Heidelberg, New York, (1970).

[17] KELLEY, J.L., NAMIOKA, I and Co-authors.: Linear topological spaces, Van Nostrand, Princeton, (1963).

[18] KELLEY, J.L.: General Topology; Van Nostrand, (1955).

[19] KHALEELULLA, S.M.: Semi-inner product spaces and algebras, Shafco Publ., Hassan-573201, India, (1981).

[20] KÖTHE, G.: Topological vector spaces I (English edition), Springer-Verlag, New York, (1969).

[21] LOOMIS, L.H.: An introduction to abstract harmonic analysis, Ven Nostrand, (1953).

[22] MARTI, J.: Introduction to the theory of bases, Springer Tracts in Natural Philosophy, Vol. 18, New York,(1969).

[23] MICHAEL, E.A.: Locally multiplicatively convex topological algebras, Mem. Amer. Math. Soc. No. 11 (Third printing), (1971).

[24] NAIMARK, M.A.: Normed rings, Noordhoff, (1959).

[25] NAMIOKA, I.: Partially ordered linear topological spaces, Mem Amer. Math. Soc. No. 24, (1957).

[25(a)] NATASON, I.P.: Constructive Theory of functions, (English edition), New York, (1964-65).

[26] PERESSINI, A.L.: Ordered topological vector spaces, Harper and Row Publ., New York, (1967).

[27] RICKART, C.E.: Banach algebras, Van Nostrand, Princeton, (1960).

[28] RIESZ, F. and Sz.-NAGY.: Functional analysis (English edition), Blackie, London, (1956).

[29] ROBERTSON, A.P. and ROBERTSON, W.J.: Topological vector spaces, Camb. Tracts in Math., Camb. Univ. Press, England, (Second edition), (1973).

[30] RUDIN, W.: Functional analysis, McGraw-Hill, Inc. (1973).

[31] SCHAEFER, H.S.: Topological vector spaces, Springer-Verlag, New York-Heidelberg-Berlin, (Third Printing), (1971).

[32] SCHWARTZ, L.: Théorie des distributions, Hermann, Paris, (1950, 1951).

[33] SINGER, I.: Bases in Banach spaces, Vol. I, Springer-Verlag, New York-Heidelberg-Berlin, (1970).

[34] TAYLOR, A.E.: Introduction to functional analysis, Wiley, New York, (1958).

[35] TREVÉS, F.: Topological vector spaces, distributions and Kernels, Acad. Press, New York, (1967).

[36] WAELBROECK, L.: Topological vector spaces and algebras, Lecture Notes in Mathematics, No. 230, Springer-Verlag, New York-Heidelberg-Berlin, (1971).

[37] WILANSKY, A.: Functional analysis, Blaisdell, New York, (1964).

[38] WILANSKY, A.: Topics in functional analysis, Lecture Notes in Mathematics, No. 45, Springer-Verlag, New York-Heidelberg-Berlin,(1967).

[39] WONG, Y.-C. and NG, K.-F.: Partially ordered topological vector spaces, Oxford Mathematical Monographs, Clarendon Press, Oxford, (1973).

[40] YOSIDA,K.:Functional analysis, Springer-Verlag, Berlin-Gottingen-Heidelberg, (1965).

[41] ZELAZKO,W.: Selected topics in topological algebras, Lecture Notes Series No. 31, Matematisk Institute, Aarhus Universitet, (1971).

[42] IVES, R.S. : Semiconvexity and locally bounded spaces, Ph.D. Thesis, University of Washington, (1957).

PAPERS

[1] ALLAN, G.R.: A spectral theory for locally convex algebras, Proc. London Math. Soc.(3), 15, (1965), 399-421.

[2] ALLAN, G.R.: On a class of locally convex algebras, Proc. London Math. Soc. (3), 17, (1967), 91-114.

[3] ARENS, R.F.: The space L^{ω} and convex topological rings, Bull, Amer. Math. Soc. 52, (1946), 931-935.

[4] ARENS, R.F.: Linear topological division algebras, Bull. Amer. Math. Soc. 53, (1947), 623-630.

[5] ARSOVE, M.G.: Similar bases and isomorphisms in Fréchet spaces, Math. Analan 135, (1958), 283-293.

[6] ARSOVE, M.G. and EDWARDS, R.E.: Generalized bases in topological linear spaces, Studia Math. 19, (1960), 95-113.

[7] BOURGIN, D.G.: Linear topological spaces, Amer. J. Math. 65, (1943), 637-659.

[8] CHILANA, A.K.: Some special operators and new classes of locally convex spaces, Proc. Camb. Phil. Soc. 71, (1972), 475-489.

[9] CHILANA, A.K. and KAUSHIK, V.: Examples in linear topological spaces, J. London Math. Soc. (2), 8, (1974), 231-238.

[10] COCHRAN, A.C., KEOWN, R. and WILLIAMS, C.R.: On a class of topological algebras, Pacific J. Math., Vol. 34, No. 1, (1970), 17-25.

[11] COLLINS, H.S.: Completeness and compactness in linear topological spaces, Trans. Amer. Math. Soc. 79, (1955), 256-280.

[12] DAY, M.M.: The spaces L^p with $0<p<1$, Bull. Amer. Math. Soc. 46, (1940), 816-823.

[13] DEAN, D.W.: Schauder decompositions in (m), Proc. Amer. Math. Soc. 18, (1967), 619-623.

[14] De WILDE, M. and HOUET, C. : On increasing sequences of absolutely convex sets in locally convex spaces, Math. Ann. 192 (1971), 257-261.

[15] DIEUDONNÉ, J.: Sur les espaces de Köthe, J. Analyse Math., 1, (1951), 81-115.

[16] DIEUDONNÉ, J.: Sur les espaces de Montel metrizable, C.R. Acad. Sci. Paris. 238 (1954), 194-195.

[17] DIEUDONNÉ, J.: On biorthogonal systems, Michigan Math. J., 2, (1954), 7-20.

[18] DIEUDONNÉ, J. and SCHWARTZ, L.: La dualite dans les espaces (F) et (LF), Ann. Inst. Fourier, Grenoble, 1, (1950), 61-101.

[19] DWORETZKY, A. and ROGERS, C.A.: Absolute and unconditional convergence in normed linear spaces, Proc. Nat. Acad. Sci. U.S.A. 36 (1950), 192-197.

[20] ELLIS, A.J.: The duality of partially ordered normed linear spaces, J. London Math. Soc. 39 (1964), 730-744.

[21] ENFLO, P.: A counterexample to the approximation problem in Banach spaces, Acta Math. 130, (1973), 309-317.

[22] GELBAUM, B.R.: Expansions in Banach spaces, Duke Math. J. 17, (1950), 187-196.

[23] GELFAND, I.M. and NAIMARK, M.A.: Normed rings with their involutions and their representations, Izvestiya Akad. Nauk S.S.S.R. Ser. Mat., 12, (1948), 445-480.

[24] GILES, J.L.: Classes of semi-inner product spaces, Trans. Amer. Math. Soc. 129, (1967), 436-446.

[25] GROTHENDIECK, A.: Sur les espaces (F) and (DF), Summa Bras. Math. 3, (1954), 57-123.

[26] HUSAIN, T.: S-spaces and the open mapping theorem, Pacific J. Math. 12, (1962), 253-271.

[27] HUSAIN, T.: Two new classes of locally convex spaces, Math. Ann., 166, (1966), 289-299.

[28] HUSAIN, T. and KHALEELULLA, S.M." Countably order-quasi-barrelled vector lattices and spaces, Mathematicae Japonica, Vol. 20, (1975), 3-15.

[29] HUSAIN, T. and KHALEELULLA, S.M.: Order-quasi-ultra-barrelled vector lattices and spaces, Periodica Math. Hungarica, Vol. 6(4), (1975), 363-371.

[30] HUSAIN, T. and KHALEELULLA, S.M.: On countably, σ- and sequentially barrelled spaces, Canad. Math. Bull. Vol. 18(3), (1975), 431-432.

[31] HYERS, D.H.: Locally bounded linear topological spaces, Rev. Ci. (lima), 41, (1939), 555-574.

[32] HYERS, D.H.: Linear topological spaces, Bull. Amer. Math. Soc. 51, (1945), 1-24.

[33] IYAHEN, S.O.: Some remarks on countably barrelled and countably quasi-barrelled spaces, Proc. Edinburgh Math. Soc. (2), 15, (1966/67), 295-296.

[34] IYAHEN, S.O.: Semiconvex spaces, Glasgow Math. J.9, (1968), 111-118.

[35] IYAHEN, S.O.: Semiconvex spaces II, Glasgow Math. J. 10, (1968), 103-105.

[36] IYAHEN, S.O.: On certain classes of linear topological spaces, Proc. London Math. Soc. 18, (1968), 285-307.

[37] IYAHEN, S.O.: On certain classes of linear topological spaces II, J. London Math. Soc. (2), 3, (1971), 609-617.

[38] JAMES, R.C.: Bases and reflexivity of Banach spaces, Ann. of Math. (2), 52, (1950), 518-527.

[39] JAMESON, G.J.O.: Topological M-spaces, Math. Z., 103, (1968), 139-150.

[40] JONES, O.T. and RETHERFORD, J.R.: On similar bases in barrelled spaces, Proc. Amer. Math. Soc. 18, (1967), 677-680.

[41] KAPLANSKY, I.: Topological rings, Amer. J. Math. 69, (1947), 153-183.

[42] KAPLANSKY, I.: Topological rings, Bull, Amer. Math. Soc. 45, (1948), 809-826.

[43] KARLIN, S.: Unconditional convergence in Banach spaces, Bull. Amer. Math. Soc. 54, (1948), 148-152.

[44] KARLIN, S.: Bases in Banach spaces, Duke Math. J. 15, (1948), 971-985.

[45] KHALEELULLA, S.M.: Countably O.Q.U. vector lattices and spaces, Tamkang J. Math. Vol. 6, No.1, (1975), 43-51.

[46] KHALEELULLA, S.M.: Countably m-barrelled algebras, Tamkang J. Math. Vol. 6, No. 2, (1975), 185-190.

[47] KHALEELULLA, S.M.: Husain spaces, Tamkang J. Math. Vol. 6, No. 2, (1975), 115-119.

[48] KLEE, V.L.: Convex sets in linear spaces III, Duke Math. J. 20 (1953), 105-111.

[49] KŌMURA, Y.: On linear topological spaces, Kumamoto J. Sc. Series A, 5, (1962), 148-157.

[50] KŌMURA, Y.: Some examples on linear topological spaces, Math. Ann. 153, (1964), 150-162.

[51] KREIN, M. and MILMAN, I.: On extreme points of regular convex sets, Stud. Math. 9, (1940), 133-138.

[52] KUCZMA, M.E.: On a problem of E. Michael concerning topological divisors of zero, Coll. Math. 19, (1968), 295-299.

[53] LANDSBERG, M.: Pseudonormen in der theorie der linearen topologischen raume, Math. Nachr. 14, (1955), 29-38.

[54] LANDSBERG, M.: Lineare topologische raume die nicht lokelkonvex sind, Math. Z., 65, (1956), 104-112.

[55] LAVERELL, W.D.: Between barrelled and W-barrelled for C(X), Indiana Univ. Math. J., Vol. 22, No.1, (1972), 25-31.

[56] LEVIN, M. and SAXON, S.: A note on the inheritance of properties of locally convex spaces by subspaces of countable codimension, Proc. Amer. Math. Soc. 29, No. 1, (1971), 97-102.

[57] LIVINGTON, A.E.: The space H^p, $0<p<1$, is not normable, Pacific J. Math. 3, (1953), 613-616.

[58] LUMER, G.: Semi-inner product spaces, Trans. Amer. Math. Soc. 100, (1961), 29-43.

[59] MACKEY, G.W.: On infinite-dimensional linear spaces, Trans. Amer. Math. Soc. 57, (1945), 155-207.

[60] MACKEY, G.W.: On convex topological linear spaces, Trans. Amer. Math. Soc. 60, (1946), 519-537.

[61] MAHOWALD, M. and GOULD, G.: Quasibarrelled locally convex spaces, Proc. Amer. Math. Soc. 2, (1960), 811-816.

[62] MALLIAVIN, P.: Impossibilité dela synthése spectrale sur groups abeliens noncompact, Faculte des Sciences de Paris, Seminaire d' analyse (P. Lelong), 1958-1959.

[63] MALLIOS, A.: On the spectra of topological algebras, J. of Functional analysis, 3, (1969), 301-309.

[64] MARKUSHEVICH, A.: Sur les bases (an sens large) dans les espaces linearies, Doklady Akad, Nauk, S.S.S.R., 41, (1943), 227-229.

[65] McARTHER, C.W. and RETHERFORD, J.R.: Uniform and equicontinuous Schauder bases of subspaces, Canad. J. Math., 17, (1965), 207-212.

[66] MORRIS, P.D. and WULBERT, D.E.: Functional representation of topological algebras, Pacific J. Math. Vol. 22, No. 2, (1967), 323-337.

[67] NACHBIN, L.: Topological vector spaces of continuous functions, Proc. Nat. Acad. Sci. Wash., 40, (1954), 471-474.

[68] NAKANO, H.: Linear topologies on semi-ordered linear spaces, J. Foc. Sci. Hokkaido Univ. 12, (1953), 87-104.

[69] NATH, B.: Topologies on generalized semi-inner product spaces, Compositio Math., Vol. 23, Fasc. 3, (1971), 309-316.

[70] NG, K.F.: Solid sets in ordered topological vector spaces, Proc. London Math. Soc. (3), 22, (1971), 106-120.

[71] PERSSON, A.: A remark on the closed graph theorem in locally convex spaces, Math. Scand., 19, (1966), 54-58.

[72] PRUGOVECKI, E.: Topologies on generalized inner product spaces, Canad, J. Math., 21, (1969), 158-169.

[73] PTÁK, V.: On complete topological linear spaces, Ceh. Math. Zur., 3(78), (1953), 301-364.

[74] ROBERTSON, W.: Completion of topological vector spaces, Proc. London Math. Soc. 8, (1958), 242-257.

[75] ROLEWICZ, R.: Example of semisimple m-convex B_0 - algebra which is not a projective limit of semi-simple B-algebras, Bull. Acad. Polon. Sci. Ser. Math. Astronem Phys., 11, (1963), 459-462.

[76] RUTHERFORD, J.R.: Two counterexamples to a conjecture of S. Karlin, Bull. Acad. Polon. Sci., 16, (1968), 293-295.

[77] SAXON, S.A.: Nuclear and product spaces, Baire-like spaces and the strongest locally convex topology, Math. Ann., 197, (1972), 87-106.

[78] SAXON, S. and LEVIN, M.: Every countable-codimensional subspace of a barrelled space is barrelled, Proc. Amer. Math. Soc., 29, No.1, (1971), 91-96.

[79] SCHAEFER, H.H.: On the completeness of topological vector lattices, Mich. Math. J., 7, (1960), 303-309.

[80] SCHAUDER, J.: Zur theorie Abbildungun in Funktionalraumen, Math. Z., 26, (1927), 47-65.

[81] SCHAUDER, J.: Eine Eigenschaft des Haarschen orthogonal system Math. Z., 28, (1928), 317-320.

[82] SCHMETS, J.: Indépendence des propriétés de tonnelage et dévaluabilité affaiblis, Bull. Soc. Roy. Sc., Liége, 42e annee, n° 3-4, (1973), 104-108.

[83] SEGAR, I.E.: The group algebra of a locally compact group. Trans. Amer. Math. Soc. 61, (1947), 69-105.

[84] SEGAR, I.E.: Irreducible representation of operator algebras, Bull. Amer. Math. Soc. 53, (1947), 73-88.

[85] SHIROTA, T.: On locally convex vector spaces of continuous functions, Proc. Jap. Acad. 30, (1954), 294-299.

[86] SIMONS, S.: Boundedness in linear topological spaces, Trans. Amer. Math. Soc.,Vol. 113, No. 1, (1964), 169-180.

[87] SINGER, I.: On the basis problem in topological linear spaces, Rev. Roumeine Math. Pures et Appl. 10, (1965), 453-457..

[88] SNIPES, R.F.: C-sequential and S-bornological topological vector spaces, Math. Ann. 202, (1973), 273-283.

[89] TWEDDLE, I.: Some remarks on σ-barrelled spaces, (unpublished).

[90] WARNER, S.: Inductive limits of normed algebras. Trans. Amer. Math. Soc. 82 (1956), 190-216.

[91] WARNER, S.: The topology of compact convergence on continuous function spaces, Duke Math. J., 25, (1958), 265-282.

[92] WEBB, J.H.: Sequential convergence in locally convex spaces, proc. Camb. Phil. Soc. 64, (1968), 341-364.

[93] WESTON, J.D.: The principle of equicontinuity for topological vector spaces, Proc. Durham Phil. Soc., 13, (1957), 1-5.

[94] WONG, Y.-C.: Order-infrabarrelled Riesz spaces, Math. Ann., 183, (1969), 17-32.

[95] ZELAZKO, W.: On the locally bounded and m-convex topological algebras, Stud. Math., 19, (1960), 333-356.

INDEX

Absolutely convergent basis 112

Absolutely convergent series 111

Absolute value 7

Absorbing 3

A-convex algebra 141

$\aleph$-hyperbarrelled space 67

$\aleph$-quasihyperbarrelled space 67

Algebra 1

Algebraic dual 20

Almost Archimedean 6

Almost open 18

Approximate identity 138

Approximate order-unit 6

Approximate order-unit normed space 81

Archimedean 6

A*-algebra 138

Auxiliary norm 138

Baire space 27

Baire-like space 27

Balanced 3

Banach algebra 138

Banach lattice 79

Banach space 18

Banach *-algebra 138

Band 8

Barrel 28

Barrelled space 28

Base for positive cone 7

Basis 111
Base norm 81
Base normed space 81
Base semi-norm 81
B-complete 18
b-cone 78
Besselian basis 113
Biorthogonal system 112
Bipolar 21
Bornivorous 28
Bornivorous suprabarrel 66
Bornivorous ultrabarrel 66
Bornivorous ultrabarrel of type (α) 66
Bornological space 29
Bounded multiplier convergent 111
Bounded set 10
Boundedly complete basis 112
Boundedly order-complete 79
(b)-Schauder basis 114
B*-algebra 138
Canonical bilinear functional 20
Cartesian product space 103
Circled 3
Circle operation 137
Closed neighbourhood condition 68
Compatible 9
Complete 10
Complex vector space 1
Cone 5
Convex 3

Coefficient functionals 111
C.O.Q. vector lattice 82
Countably barrelled space 33
Countably m-barrelled algebra 140
Countably O.Q.U. vector lattice 84
Countably quasibarrelled space 33
Countably quasi-ultrabarrelled space 66
Countably ultrabarrelled space 66
C-sequential space 30
Decomposable 6
Decomposition property 7
(DF) space 33
Dimension 2
Direct sum 2
Directed set 5
Distinguished space 32
Dual 20
E-complete biorthogonal system 112
(e)-Schauder basis 114
Exhausting 6
Extended Markushevich basis 114
Extremal point 7
Filter condition 68
Fréchet algebra 138
Fréchet space 18
Full 6
GB*-algebra 141
Gelfand map 139
Generalized basis 114

Generalized inner product space 18

Generalized semi-inner product space 18

Generating cone 6

Hamel basis 2

Hilbertian basis 113

Hilbert space 18

H-space 33

Hyperbarrelled space 67

Hyperbornological space 67

Hyper-quasibarrelled space 67

Idempotent 137

Identity 138

Inductive limit 28

Infimum 7

Inner product 4

Inner produce space 18

Isomorphism Theorem for bases 115

k-barrelled space 33

k-norm 3

k-quasibarrelled space 33

k-quasiultrabarrelled space 67

k-semi-norm 3

k-ultrabarrelled space 67

Lattice homomorphism 8

Lattice ideal 8

Lattice operations 8, 82

(LB)-space 28

Left ideal 137

Left topological divisor of zero 139

(LF)-space 28
Linear combination 2
Linear functional 3
Linear map 3
Linearly independent 2
Locally bounded space 11
Locally convex algebra 138
Locally convex lattice 79
Locally convex space 12
Locally convex *-algebra 140
Locally m-convex algebra 138
Locally m-semiconvex algebra 142
Locally semi-convex space 11
Locally solid topology 79
Locally topological space 55
L-W space 65
Mackey space 31
Majorized 5
Markushevich basis 114
Maximal biorthogonal system 113
Maximal ideal 137
m-barrelled space 140
m-bounded 139
m-convex 137
Metrizable 10
Minorized 5
Modular ideal 137
Monotone basis 112
Montel space 32
$\mathcal{M}$-singular 139

M-space 82
Negative part 7
Norm 3
Normal basis 112
Normal cone 77
Normal topology 47
Normed algebra 137
Normed space 18
Normed vector lattice 79
Nowhere dense 27
N-S space 65
ω-linearly independent 122
Open decomposition 78
O.Q.U. vector lattice 84
Order-bornivorous 82
Order-bornivorous ultrabarrel 83
Order-bornivorous ultrabarrel of type (α) 84
Order-bound topology 81
Order-bounded 5
Order-complete 7
Order-continuous 8
Order-convergence 8
Order-(DF) vector lattice 83
Ordered locally convex space 77
Ordered topological vector space 77
Ordered vector space 5
Order-interval 5
Order-quasibarrelled space 82
Order-separable 8
Order-unit 6

Order-unit normed space 81

P-algebra 139

p.i.b. algebra 139

p-normed algebra 141

Polar 21

Positive cone 5

Positive linear functional 7

Positive linear map 7

Positive part 7

Precompact 10

Pre-Hilbert space 18

Product 2

Property (C) 31

Property (S) 31

Pseudo-complete 140

Pseudo-M-space 82

Q-algebra 139

Q-space 29

Quasi-barrelled space 28

Quasi-complete 10

Quasi-hyperbarrelled space 67

Quasi-M-barrelled space 31

Quasi-norm 4

Quasi-regular 137

Quasi-semi-norm 3

Quasi-ultrabarrelled space 66

Quotient map 103

Quotient space 2,103

Quotient topology 103

Radical 137

Real-compact	29
Real vector space	1
Reflexive	31
Regular ideal	137
Retro-basis	113
Right ideal	137
Right topological divisor of zero	139
S-bornological space	30
Schauder basis	111
Schwartz space	32
Semi-bornological space	30
Semi-complete	10
Semi-convex	3
Semi-inner product	4
Semi-inner product space	18
Semi-norm	3
Semi-reflexive	31
Semi-simple	137
Sequentially barrelled space	33
Sequentially complete	10
Sequentially order-continuous	8
σ-barrelled space	33
σ-order complete	7
σ-quasibarrelled space	33
Similar bases	115
Singular	139
Solid set	8
Solid ultrabarrel	83
Solid ultrabarrel of type (α)	84
Sublattice	8

Subseries convergent 111
Subspace 1
Subsymmetric basis 113
Suprabarrel 65
Supremum 7
$\mathfrak{S}$-topology 19
Strict b-cone 78
Strict inductive limit 28
Strong topology 20
Symmetric Banach *-algebra 138
Symmetric locally convex *-algebra 140
Symmetric basis 113
Topological divisor of zero 139
Topological dual 20
Topological vector lattice 79
Topological vector space 9
t-polar space 28
Topology of compact convergence 19
Topology of precompact convergence 19
Topology of simple convergence 19
Topology of uniform convergence on bounded sets 19
Total 19
Totally bounded 10
Two-sided ideal 137
Ultrabarrel 65
Ultrabarrel of type (α) 66
Ultrabarrelled space 66
Ultrabornological space 66
Unconditional basis 112
Unconditional convergent series 110

Unit vector basis	120
Unordered Baire-like space	27
Unordered convergent series	110
Vector space	1
Vector lattice	7
W-barrelled space	65
Weak basis	111
Weak basis theorem	112
Weak topology	20
Weak* topology	20
Wedge	5
Wiener property	147